Herbert Edward Cox

A Handbook of the Coleoptera

Beetles of Great Britain and Ireland - Vol. 2

Herbert Edward Cox

A Handbook of the Coleoptera
Beetles of Great Britain and Ireland - Vol. 2

ISBN/EAN: 9783337144036

Printed in Europe, USA, Canada, Australia, Japan

Cover: Foto ©berggeist007 / pixelio.de

More available books at **www.hansebooks.com**

A

HANDBOOK

OF THE

COLEOPTERA

Or Beetles,

OF

GREAT BRITAIN AND IRELAND,

BY

HERBERT E. COX, M.E.S.

VOL. II.

London:

E. W. JANSON, 28, MUSEUM STREET.

1874.

NORMAL COLEOPTERA

(CONTINUED).

LAMELLICORNIA.

A. Club of antennae comblike, plates fixed.

Antennae with ten joints. Epimera of metathorax covered; abdomen with five ventral segments; spiracles placed in articulating membrane between dorsal and ventral segments. First four tarsal joints about equal in length, last joint with a prominence between claws, bearing two bristles. *Lucanidae.*

B. Club of antennae formed of moveable plates.

a. Spiracles placed in articulating membrane between dorsal and ventral segments of abdomen.

Ligula separate from mentum, rarely horny. Penultimate dorsal and ventral abdominal segments joined by a membrane. Club of antennae usually three-jointed, closely covered with pubescence within.

I. Abdomen with six ventral segments.

1. Epimera of metathorax hidden; antennae with eight or nine joints.

A A. Ventral segments of abdomen soldered together; posterior tibiae with only one apical spine.

Intermediate legs usually placed far apart. Eyes partly or wholly divided; antennae rarely with eight joints; scutellum hidden or very small; apex of abdomen not covered by elytra; anterior tarsi sometimes extremely small or absent. *Copridae.*

B B. Ventral segments of abdomen freely articulated ; posterior tibiae with two apical spines.

Intermediate legs placed near each other. Eyes round
or only slightly narrowed; antennae with nine joints;
scutellum distinct; anterior tarsi distinct.   *Aphodiidae.*

2. Epimera of metathorax free ; antennae with
eleven joints.

Intermediate legs usually placed near each other. Eyes
wholly or partly divided; scutellum usually moderately
large ; apex of abdomen often covered by elytra. Ventral
segments of abdomen freely articulated; posterior tibiae
with two apical spines.   *Geotrupidae.*

II. Abdomen with five ventral segments.

Intermediate legs placed near each other. Eyes sometimes partly divided ; scutellum rather small; apex of
abdomen covered by elytra. Epimera of metathorax
hidden ; ventral segments of abdomen slightly moveable ;
posterior tibiae with two apical spines. Antennae with
ten joints.   *Trogidae.*

b. Second to sixth pairs of spiracles placed in ventral
abdominal segments.

Ligula soldered to mentum. Abdomen with six ventral
segments, penultimate dorsal and ventral segments soldered
together. Club of antennae with from three to seven
joints, inner-side of plates bare ; last pair of spiracles uncovered.

I. Penultimate spiracles placed in the same line with
the front ones.

Clypeus generally distinct from forehead ; labrum usually
prominent, horny; ligula sometimes membranous. Antennae
with from seven to ten joints, club often stronger in male.
                                      Melolonthidae.

II. Penultimate spiracles placed more outward than
front ones.

1. Tarsal claws unequal.

Clypeus generally distinct from the forehead ; labrum
usually prominent ; ligula horny. Antennae with nine or
ten joints, club three-jointed, alike in both sexes.
                                      Rutelidae.

2. Tarsal claws equal.

Clypeus not separate from forehead; labrum hidden; mandibles hidden, membranous part much developed; ligula horny. Antennae with ten joints, club three-jointed, generally alike in both sexes. Elytra not reflexed at sides.

Cetoniidae.

LUCANIDAE.

A. Eyes partly divided.

a. Labrum deflexed; eyes divided about half through.

Outer lobe of maxillae much longer than inner, latter without horny hook in either sex; ligula split; mentum broad. Male with head large and mandibles very strongly developed. *Lucanus*, Lin.

b. Labrum prominent; eyes almost entirely divided.

Lobes of maxillae about equal in length, inner one with a horny hook in female; ligula split; mentum broad.

Dorcus, Mc.L.

B. Eyes entire.

Labrum hidden; maxillae with only one lobe; ligula entire, small; mentum narrow. *Sinodendron*, Hellw.

Lucanus.

Black, elytra chestnut-brown. Club of antennae with four plates; posterior tibiae with three teeth. Male with head as broad as, or broader than thorax; mandibles as long or nearly as long as elytra, strongly toothed. Female with head much narrower than thorax; mandibles small, with two rather blunt teeth in middle. L. (without mandibles) 1—2 inches. Common. *L. cervus*, Lin.

Dorcus.

Black, dull. Elytra very closely punctured, slightly wrinkled; posterior tibiae with a little tooth near middle. Male with head as broad as thorax, latter finely and rather diffusely punctured, labrum truncate and mandibles shorter than head. Female with head narrower than thorax, latter very closely punctured; labrum emarginate and mandibles small. L. 7—10 l. Common. *D. parallelopipedus*, Lin.

B 2

Sinodendron.

Black, shiny. Elytra roughly punctured, with indistinct striae. Male with a long curved horn on head, front of thorax cut off obliquely, with five teeth. L. 4½—6 l. Common.   *S. cylindricum*, Lin.

COPRIDAE.

A. Labial palpi with first joint much longer than second, third joint distinct.

Anterior tarsi present; joints of posterior tarsi gradually diminishing in length and breadth.   *Copris*, Geoffr.

B. Labial palpi with first joint shorter than second; third extremely small.

Anterior tarsi present; first joint of posterior tarsi elongate.   *Onthophagus*, Latr.

Copris.

Black, shiny. Head with a horn (long in male, short in female); thorax sloped in front, strongly bordered at base, with a central furrow; elytra with notched striae, interstices slightly arched; last dorsal segment of abdomen rather closely punctured. L. 7—10 l. Rather common.

C. lunaris, Lin.

Onthophagus.

A. Head with one or two horns in male; length 2½—5 lines.

a. Elytra unicolorous black.

I. Thorax rather diffusely and finely punctured; male with two long curved horns on head.

Black, dull; thorax usually slightly greenish. Head somewhat oblong, rounded in front; elytra with feeble punctured striae; interstices even, diffusely punctured. Female with two straight, transverse ridges on forehead. L. 3½—5 l. Rare.   *O. taurus*, Lin.

II. Thorax closely punctured, granulate; male with vertex of head prolonged into a plate, narrowing into a horn, bent forward.

Black, dull, thorax sometimes greenish. Elytra with

feeble punctured striae, interstices finely granulate. Female with two transverse ridges (front one slightly bent) on forehead, front of head very slightly emarginate and front of thorax with a small double prominence. L. 4 l. Not common. *O. nutans*, Fab.

b. Elytra yellowish, more or less sprinkled with brown or black.

1. Thorax rather closely granulated.

Head and thorax bronze-green ; elytra reddish-yellow, with greenish-brown markings. Clypeus slightly emarginate in front ; sides of thorax evenly rounded ; elytra with feeble punctured striae, interstices slightly convex, finely and rather diffusely granulate, almost in rows. Male with vertex of head prolonged into a plate, ending in a thin upright horn ; thorax sloped in front, emarginate in middle. Female with two transverse ridges on forehead, the front one slightly bent, the hinder one with a prominence at each end ; front of thorax with a small double prominence. L. 3¾—5 l. Rather common. *O. vacca*, Lin.

II. Thorax closely punctured, not granulate.

1. Sides of thorax slightly sinuate in front, anterior angles acute.

Male with vertex of head prolonged into a plate, narrowing into a horn, bent forward. Female with two transverse ridges on forehead, the front one slightly bent.

A A. Elytra dull reddish-yellow, indistinctly sprinkled with brown or brownish-black.

Body broader and more convex than *O. fracticornis,* rather shiny. Head and thorax coppery-reddish, sometimes greenish ; antennae red, with blackish-club. Head oblong, slightly emarginate in front ; thorax very closely punctured ; elytra with feeble punctured striae, interstices diffusely punctured. Female with a slight prominence in front of thorax. L. 3—4 l. Moderately common.

O. coenobita, Herbst.

B B. Elytra yellow, sprinkled with black.

Rather dull. Head and thorax dark bronze ; antennae brown, with black club. Head oblong, slightly emarginate in front ; thorax closely punctured ; elytra with feeble punctured striae, interstices punctured almost in two rows.

Female without prominence on front of thorax. L. 2½—4 l.
Common.   *O. fracticornis*, Preys.

2. Sides of thorax rounded throughout; anterior angles blunt.

Head and thorax black, with slight bronze reflection; elytra yellow, sprinkled with a network of black. Head rounded, slightly emarginate in front; thorax punctured closely in front, gradually more diffusely behind; elytra with very feeble punctured striae, interstices flat, diffusely and finely granulate. Male with a thin upright horn on forehead, vertex of head not produced into a plate; female with two strong transverse ridges on forehead, thorax with a prominence in front. L. 2½—4 l. Common.
           O. nuchicornis, Lin.

B. Forehead with a straight transverse ridge in male; length 2—2¼ lines.

Dull black, head and thorax often slightly bronze; upper-side with short gray bristles. Head rounded, slightly emarginate in front; thorax finely punctured and granulate; elytra with very feeble notched striae, interstices flat diffusely punctured and granulate. Female with two transverse ridges, the front one bent. Common. *O. ovatus*, Lin.

APHODIIDAE.

A. Labrum and mandibles entirely hidden in mouth.

  a. Head flat; at least part of eyes uncovered when head is retracted.

Molar tooth of mandibles composed of horny plates; lobes of maxillae membranous. Posterior coxae somewhat dilated; posterior tibiae with a ring of bristles at apex.
           Aphodius, Ill.

  b. Head convex; eyes wholly hidden when head is retracted.

Molar tooth on mandibles solid.

  I. Elytra entirely covering abdomen.

Thorax without transverse furrows. Lobes of maxillae membranous. Posterior coxae as in *Aphodius*.
           Ammoecius, Muls.

  II. Half of last abdominal segment not covered by elytra.

1. Outer lobe of maxillae horny, with four teeth at apex; second joint of maxillary palpi about equal in length to third.

Head granulate, thorax with transverse furrows.

Psammodius, Latr.

2. Outer lobe of maxillae leathery, rounded ; second joint of maxillary palpi nearly double as long as third.

Head granulate ; thorax generally with four transverse furrows. *Rhyssemus,* Muls.

B. Labrum and mandibles prominent.

Head granulate, convex. Mandibles horny. Molar tooth solid, inner lobe of maxillae horny, hooked at apex, outer lobe leathery ; eyes covered when head is retracted.

Aegialia, Latr.

Aphodius.

A. Scutellum very large, $\frac{1}{5}$ or $\frac{1}{4}$ of the length of elytra.

a. Body flat.

I. Elytra dirty brownish-yellow, often darker in middle.

Head and thorax black ; antennae and palpi black. Thorax rather closely and finely punctured ; elytra with fine shallow notched striae, interstices flat, rather closely punctured. Ring at apex of hinder tibiae formed of longer and shorter bristles. L. $3\frac{1}{2}$— 4 l. Common.

A. erraticus, Lin.

II. Upper-side entirely black.

Antennae brownish-yellow, club blackish ; palpi reddish-brown. Thorax diffusely covered with large, mixed with small punctures ; elytra with strong notched striae, inner interstices raised. Ring at apex of hinder tibiae formed of short bristles only. L. $2\frac{1}{2}$—3 l. Common.

A. subterraneus, Lin.

b. Body convex.

Ring at apex of hinder tibiae formed of short bristles only.

I. Outer edge of anterior tibiae not notched above teeth ; length 4—$5\frac{1}{2}$ lines.

Oblong. Black. Elytra with rather feeble notched striae. Common.   *A. fossor*, Lin.

II. Outer edge of anterior tibiae notched above teeth; length 1½—2¼ lines.

Short. Black, apex of elytra red. Elytra with strong, closely notched striae. Common.

           A. haemorrhoidalis, Lin.

B. Scutellum small.

    a. Ring at apex of hinder tibiae formed of short bristles only.

       I. Base of thorax bordered throughout.

          1. Thorax slightly emarginate at posterior angles.

            A A. Thorax unicolorous black.

Convex, a little shorter than *A. fimetarius*. Head black, elytra grayish-yellow, sometimes darker in middle; abdomen black; antennae brownish-yellow, club blackish; legs lighter or darker brown. L. 2½—3½ l. Common.

           A. scybalarius, Fab.

            B B. Thorax black, with red spot in anterior angles.

              a a. Abdomen red.

Broader than *A. fimetarius*, with longer thorax and shorter elytra, rather flat on disc. Head black; elytra red, sometimes with a darker cloud behind; antennae red, club rust-yellow; legs black, tibiae and tarsi lighter. L. 2½—3 l. Common.   *A. foetens*, Fab.

              b b. Abdomen black.

Oblong, convex. Head black; elytra red; antennae red, club rust-yellow; legs black, tarsi lighter. L. 2½—3½ l. Common.   *A. fimetarius*, Lin.

          2. Thorax not emarginate at posterior angles.

            A A. Ground colour black.

              a a. Mesosternum finely ridged between intermediate coxae.

                A a. Thorax closely punctured.

                  A 1. Elytra unicolorous black, dull.

Short, very convex. Black; antennae brown; legs black, tarsi lighter. Clypeus with a more or less prominent transverse wrinkle in middle. L. 1¾—2½ l. Common.

A. ater, De G.

B 1. Elytra black-brown, with apex brown-red, shiny.

Oblong, convex. Head and thorax black ; antennae dark brown, club blackish ; legs black or brown, tarsi lighter. Clypeus with a raised curved transverse line in middle. Broader than *A. granarius*, with prominences on forehead stronger and scutellum larger. L. 2—2½ l. Moderately common. *A. constans*, Duft.

B b. Thorax diffusely punctured.

Oblong, slightly convex. Black ; elytra with apex of outer margin red-brown ; antennae brownish-yellow, club blackish ; legs lighter or darker red-brown. L. 1½—2½ l. Common. *A. granarius*, Lin.

b b. Mesosternum not ridged.

A a. Thorax black, with anterior angles or sides brown-red or pitch-brown.

A 1. Outer spur of posterior tibiae scarcely shorter than first joint of tarsi.

Oblong, rather convex, shiny. Black ; anterior angles of thorax pitch-brown ; elytra red ; legs pitch-brown. Clypeus somewhat roughened with punctures in front ; thorax very closely and finely punctured ; elytra with fine notched striae, interstices very finely punctured ; forehead with three tubercles, central one nearly acute in male. L. 2—2½ l. Rather common. *A. lapponum*, Gyll.

B 1. Outer spur of posterior tibiae distinctly shorter than first joint of tarsi.

a 1. Elytra dark red, sometimes darker behind.

Reversed ovate, convex. Head black, antennae red-brown, club blackish ; palpi black ; legs red-brown, femora darker ; middle joints of posterior tarsi diminishing in length. Thorax very closely punctured. Similar to *A. lapponum* but with clypeus not roughened in front, outer spur of posterior tibiae distinctly shorter than first-joint of tarsi, frontal tubercles very indistinct in male, absent in female. L. 1½—2¼ l. Scarce. *A. foetidus*, Fab.

b 1. Elytra smoke-brown, with red
spots at shoulder, scutellum and
before apex.

Reversed ovate, very convex. Head black, rather dull;
antennae brownish-red, club sometimes darker; palpi red-
brown; legs brown-red; middle joints of posterior tarsi
about equal in length. Thorax rather closely punctured.
L. 1½ l. Rather common.   *A. putridus*, Creutz.

B b. Thorax unicolorous black.

Reversed ovate, slightly convex. Head and elytra black;
antennae, palpi and legs red-brown, club of former and
femora blackish. Clypeus with a raised, slightly curved
transverse line, prominences on forehead distinct, middle
one not higher than side ones in male, thorax moderately
finely and evenly punctured; middle joints of posterior
tarsi diminishing in length. L. 2—2¼ l. Scarce.
          A. nemoralis, Er.

B B. Ground colour brown-red or yellowish.

a a. Mesosternum not ridged.

A a. Elytra yellowish, with a narrow brown
sutural streak; legs reddish-yellow, femora
light yellow.

Oblong. Head yellowish-red-brown; darker in middle;
thorax with disc blackish, margins yellow; breast brown,
with a triangular yellow spot on each side; abdomen
yellow. Thorax diffusely and very finely punctured;
elytra with very short pubescence toward apex. L. 2½—
3½ l. Common.   *A. sordidus*, Fab.

B b. Elytra and legs rust-red or brown-red.

Rust-red or brown-red, disc of thorax and breast brown.
Rather more convex than *A. sordidus*, more steeply sloped
behind, head even, thorax slightly narrower and more closely
punctured, elytra without pubescence behind, apical margin
closely punctured and shiny. L. 2½—3 l. Common.
          A. rufescens, Fab.

b b. Mesosternum finely ridged between intermediate coxae.

Elongate, nearly cylindrical, very shiny. Head red-
brown, with blackish vertex; thorax black, with sides
yellowish-brown-red; elytra reddish-yellow, suture nar-

rowly brownish ; under-side brown-red, breast with a light central spot. L. 2—2½ l. Not uncommon.

A. nitidulus, Fab.

II. Base of thorax bordered at angles only.

1. Mesosternum finely ridged between intermediate coxae.

A A. Outer spur of posterior tibiae longer than first joint of tarsi.

Elongate, nearly cylindrical. Black, generally metallic ; elytra sometimes with a red spot. Thorax extremely finely punctured, with an admixture (scanty on disc, closer toward sides) of larger punctures ; metasternum rather finely and closely punctured, in male hairy in middle, and sometimes depressed. L. 1½—2 l. Not very common.

A. plagiatus, Lin.

B B. Outer spur of posterior tibiae shorter than first joint of tarsi.

Similar to *A. plagiatus*, but with head more closely punctured, cheeks less prominent, the larger punctures on thorax not so large, and metasternum finely and diffusely punctured, with the depression not hairy in male. L. 2 l. Rare. *A. niger*, Panz.

2. Mesosternum not ridged.

Oblong, convex, very shiny. Pale brownish-yellow, base of head, disc of thorax, suture and an oblong spot on each elytron brown. Prominences on forehead distinct, middle one high. L. 1½—2¼ l. Not common. *A. lividus*, Ol.

b. Ring at apex of hinder tibiae formed of longer and shorter bristles.

I. Prominences on forehead distinct (especially in male).

1. Base of thorax bordered.

A A. Mesosternum finely ridged between intermediate coxae.

Oblong, rather convex, shiny. Black ; anterior angles of thorax sometimes red-brown ; elytra yellow, suture, three or four spots and a lateral streak black ; antennae and palpi black-brown ; legs red-brown, femora yellow beneath. Elytra with extremely fine, short hairs toward apex. L. 1½—3 l. Common. *A. inquinatus*, Fab.

B B. Mesosternum not ridged.

a a. Femora yellow, at least beneath.

A a. Palpi yellow ; femora entirely yellow.

Oblong, slightly convex. Black; head with a brown-red spot on each side ; sides of thorax yellow ; scutellum brown, with a yellow spot in middle; elytra gray-yellow, with two narrow bands of black spots and with the striae brown ; metasternum yellow in middle ; antennae yellow, with brownish club. Elytra with extremely fine, short hairs toward apex. L. 1¾—2¾ l. Moderately common.

A. sticticus, Panz.

B b. Palpi black or dark brown; femora yellow beneath only.

Oblong, convex. Black; head with a red-brown spot on each side; sides of thorax reddish-yellow; elytra yellow, each with seven small oblong black spots ; metasternum brownish-yellow in middle ; antennae brownish-yellow with brown club. Elytra without hairs. L. 2—2½ l. Not common. *A. conspurcatus,* Lin.

b b. Femora brown.

Much shorter and more convex than *A. inquinatus.* Black ; sides of thorax sometimes dark red-brown ; elytra brownish-yellow, with suture and two bands, formed partly of square spots, black; antennae brown, club blackish; palpi black-brown. Elytra without hairs. Male with posterior femora dilated. L. 1½—2¼ l. Not common.

A. tessulatus, Payk.

2. Base of thorax not bordered.

Oblong, convex. Brown-red ; head and disc of thorax black; elytra with an indistinct black spot behind middle ; antennae and palpi yellow ; legs brown-red. Elytra with broad, shallow notched striae, interstices raised into a ridge in middle and each with a row of large, somewhat confluent punctures. L. 2 l. Not common. *A. Zenkeri,* Germ.

II. Prominences on forehead indistinct or absent.

1. Base of thorax bordered.

A A. Posterior angles of thorax obtuse.

a a. Interstices on elytra wrinkled, dull.

A a. Mesosternum not ridged.

Oblong, rather flat. Black, dull; elytra dark-red; antennae yellow, club gray; palpi red-yellow. Forehead with three prominences; elytra with strong striae, interstices almost groove-like. L. 2—2½ l. Not very common.

A. porcus, Fab.

B b. Mesosternum with traces of a ridge between intermediate coxae.

Oblong oval, rather flat. Black, dull; sides of thorax and elytra often red-brown; antennae and palpi brown, club of former black. Thorax and elytra covered with short yellowish-gray pubescence, latter with strong striae, interstices flat; forehead without prominences. L. 1—1½ l. Rare.

A. scrofa, Fab.

b b. Interstices on elytra finely punctured.

A a. Mesosternum finely ridged between intermediate coxae.

A 1. Elytra black or blackish, with red spots.

a 1. Clypeus raised into a slight prominence in middle.

A 2. Thorax unicolorous black.

Reversed ovate, slightly convex. Black; elytra sometimes brownish-black, with red markings at shoulder and before apex. Elytra with strong notched striae, interstices almost groove-like; first joint of posterior tarsi half as long again as second. Posterior tibiae of male strongly compressed and dilated. L. 1¾—2¼ l. Not very common.

A. tristis, Panz.

B 2. Thorax black, with brown-red spot in anterior angles.

Reversed ovate, slightly convex. Black; elytra with apex red-brown, sometimes entirely lighter. Elytra with deep notched striae, interstices slightly arched; first joint of posterior tarsi almost as long as second and third together. L. 1½—2 l. Common.

A. pusillus, Herbst.

b 1. Clypeus with a slight depression on each side and in front.

Oblong, slightly convex. Black; elytra with a red spot near shoulder and another near apex of each, apex of outer margin reddish-brown. Elytra with notched striae, interstices flat and broad; first joint of posterior tarsi about as long as second and third together. L. 1¼—1¾ l. Moderately common.   *A. quadrimaculatus*, Lin.

B 1. Elytra yellowish, with suture brown.

Oblong, slightly convex. Head and thorax black, anterior angles or sides of latter yellow. Elytra with notched striae, interstices slightly arched; first joint of posterior tarsi about as long as the next three together. L. 1½—2 l. Common.   *A. merdarius*, Fab.

B b. Mesosternum not ridged.

     A 1. Head black, with a yellow spot on each side.

Oblong, slightly convex. Thorax black, sides yellow; elytra gray-yellow, each with a brown, longitudinal spot; under-side black, middle of metasternum and apex of abdomen yellow. Elytra with fine notched striae; apical spines of anterior tibiae straight and pointed in both sexes. L. 1½—2½ l. Rare.   *A. consputus*, Cr.

     B 1. Head unicolorous black.

         a 1. Apical spines of anterior tibiae dissimilar in male and female.

Oblong, slightly convex. Black; sides of thorax and middle of metasternum yellow; elytra gray-yellow, each with a brownish spot, narrow in front, dilated behind; abdomen brown, with yellowish apex. Apical spines of anterior tibiae with blunt toothed apex in male, pointed in female. L. 2—3½ l. Common.   *A. prodromus*, Brahm.

         b 1. Apical spines of anterior tibiae pointed in both sexes.

Very similar to *A. prodromus* but with traces of prominences on forehead. Male shorter and flatter than female. L. 2—3 l. Common.   *A. punctatosulcatus*, Sturm.

     B B. Posterior angles of thorax rounded.

         a a. Sides of thorax fringed with long hairs.

Oblong, slightly convex. Black, with bronze reflection, sides of head and thorax yellow; elytra grayish-yellow, with black spots, rather closely covered with gray pubescence, almost in rows; apex of abdomen yellow. Apical spines of anterior tibiae in male rather thick, cut off obliquely at apex. L. 2½—3¼ l. Common.

A. contaminatus, Herbst.

b b. Sides of thorax not fringed.

Oblong, convex. Black, with bronze reflection ; sides of head and thorax yellow, elytra grayish-yellow, with black spots, more finely pubescent than A. contaminatus; apex of abdomen yellow. Male with apical spines of anterior tibiae gradually pointed. L. 2—2½ l. Not very common.

A. obliteratus, Panz.

2. Base of thorax not bordered.

A A. Elytra with notched or punctured striae.

a a. Elytra with sixth and eighth interstices separate and apex rounded.

A a. Length 5—6 lines.

Oblong, rather convex. Upper-side dark brown, forehead and disc of thorax black ; under-side and legs red-brown ; antennae and palpi brown-red. Thorax diffusely and extremely finely punctured, with larger punctures toward anterior angles. Common. A. rufipes, Lin.

B b. Length 3—4½ lines.

A 1. Palpi black ; apex of elytra with fine pubescence.

Oblong oval, rather flat. Black ; elytra black or brownish-yellow, with black markings ; tarsi reddish. Thorax rather closely and finely punctured ; interstices on elytra diffusely punctured. Common. A. luridus, Fab.

B 1. Palpi red ; apex of elytra bare.

Nearly oval. Black ; elytra black, red with black markings, or entirely red ; tarsi reddish. Thorax finely and rather diffusely punctured with an admixture of larger punctures ; interstices on elytra rather closely punctured. Body more evenly arched than in A. luridus. Not uncommon. A. depressus, Kug.

b b. Elytra with sixth and eighth interstices united behind and apex produced into a little tooth.

Oblong, moderately convex. Black or pitch-brown, elytra often lighter ; antennae and palpi reddish-yellow ; legs brown-red. Elytra with notched striae, deeper behind ;

interstices diffusely punctured. L. 1—1½ l. Moderately common.   *A. arenarius*, Ol.

B B. Elytra with six raised longitudinal lines on each.

a a. Elytra yellowish, with second and fourth ridges checkered with black.

Oblong, rather flat, pubescent. Head and thorax red-brown; antennae and palpi yellow; legs brown-red, femora yellow. Anterior tibiae of male somewhat long, dilated and emarginate. L. 1½—2¼ l. Not uncommon.

                                  A. sus, Herbst.

b b. Elytra red-brown, usually blackish on disc, with round, yellowish spots.

Oblong, rather flat, pubescent. Head and thorax black; antennae and palpi red-brown, former with black club; legs reddish-brown. First joint of posterior tarsi almost as long as next three together. L. 1½—2 l. Rather common.

                            A. testudinarius, Fab.

c c. Elytra unicolorous brown.

Oblong, rather flat, pubescent. Brown; antennae and palpi yellow; legs light brownish-red. First joint of posterior tarsi as long as second and third together. L. 1½—2 l. Rare.   *A. villosus*, Gyll.

C C. Elytra with ten raised longitudinal lines on each.

Oblong, rather flat and dull. Brownish-black; antennae and palpi reddish-yellow; apex of abdomen and legs red-brown. L. 1—1¼ l. Rather common. *A. porcatus*, Fab.

Ammoecius.

Very short, strongly convex. Black, shiny. Clypeus with a transverse wrinkle in front, angles obtuse, almost rounded; thorax diffusely punctured at sides; elytra short, with strong notched striae. L. 1¾—2¼ l. Rare.

                                 A. brevis, Er.

Psammodius.

A. Tarsal claws small and weak; thorax bordered with bristles.

Reversed ovate, convex. Pitch-brown or brown. Thorax

with five coarsely punctured transverse furrows; first joint of posterior tarsi dilated, shorter than the broad apical spine of tibia; posterior femora thickened. L. 1½ l. Not very common. *P. sulcicollis*, Ill.

B. Tarsal claws of usual size; thorax not bordered with bristles.

Elongate, nearly cylindrical. Black; margins of thorax and usually also the elytra lighter. Thorax diffusely and very strongly punctured, with a fine central furrow, abbreviated in front, and two feeble transverse furrows on each side; first joint of posterior tarsi elongate, parallel-sided, longer than apical spine of tibiae, which is narrow and pointed. L. 1¼—1½ l. Rare. *P. caesus*, Panz.

Rhyssemus.

Brown-black, dull. Forehead diffusely granulate, vertex somewhat depressed on each side, very closely and finely granulate; thorax with four transverse furrows, the hinder two interrupted in middle, sides and base fringed with bristles; elytra with punctured striae, interstices with two rows of granules. L. 1¼—1½ l. Rare. *R. germanus*, Lin.

Aegialia.

A. Posterior legs not thickened; apical spines of tibiae narrow and pointed; tarsal claws of usual length but thin.

Oblong, almost cylindrical, winged. Black or pitch-brown. Last joint of maxillary palpi thin, spindle-shaped; thorax strongly punctured, bordered at base; elytra with strong punctured striae. L. 2¼ l. Not uncommon.
A. sabuleti, Payk.

B. Posterior legs thickened, with apical spines of tibiae dilated; tarsal claws small and feeble.

a. Body short, black or pitch-brown, elytra almost globular, with feeble striae.

Apterous. Thorax smooth, base not bordered. L. 2—2½ l. Common. *A. arenaria*, Fab.

b. Body almost cylindrical, brown-red; elytra with strong striae.

Winged. Thorax wrinkled in all directions, bordered at base. L. 2¼ l. Rare. *A. rufa*, Fab.

GEOTRUPIDAE.

A. Anterior femora with a row of hairs on front.

Club of antennae with third joint smaller than and received by second, and this by the larger first ; lobes of maxillae horny, inner one armed with two hooks, outer one rounded. *Odontaeus*, Klug.

B. Anterior femora with a spot of hairs on front.

Club of antennae with joints either entirely free or with middle joint partly enclosed by the others ; lobes of maxillae leathery, rounded, pubescent. *Geotrupes*, Latr.

Odontaeus.

Short ovate. Upper-side black or brown, sometimes lighter ; under-side brownish-yellow. Elytra with strong punctured striae ; anterior tibiae with eight teeth. Male with a long, thin, gently curved, moveable horn on head ; thorax with a horn and depression on each side, and two teeth in middle. Female with two slight elevations on forehead, and three prominences in front of thorax. L. 3 —4 l. Rare. *O. mobilicornis*, Fab.

Geotrupes.

A. Club of antennae with plates free ; thorax of male with three horns in front.

Black, shiny, rather flat. Clypeus with an indistinct longitudinal ridge in middle, forehead impressed in middle. L. 7—9 l. Common. *G. Typhoeus*, Lin.

B. Club of antennae with middle plate half enclosed ; front of thorax without horns.

   a. Body oblong oval ; hinder tibiae with three transverse ridges on outer-side.

      I. Elytra with seven striae on each between suture and humeral prominence.

         1. Mandibles twice sinuate on inner-side, straight on outer-side.

Upper-side usually dull black, with margins of thorax

and elytra bluish ; under-side violet, shiny. Disc of thorax diffusely and finely punctured ; segments of abdomen almost hairless in middle beneath. Male with third tooth from apex of anterior tibiae bent downward ; posterior femora with a large tooth on under-side and a second, smaller one formed by apex of trochanters. L. 7—11 l. Common. *G. mesolcius*, Th. (*stercorarius*, Sharp's Cat.)

> **2.** Mandibles once sinuate on inner-side, outer-side rounded.

Shiny. Upper-side blue-black ; under-side steel-blue. Rather more oblong than *G. mesolcius*, with disc of thorax smooth and striae on elytra not so well defined, with more arched interstices. Segments of abdomen evenly hairy throughout beneath. Male with two small, equal teeth on posterior femora and a smooth ridge on hinder side of anterior tibiae. L. 8—11 l. Common.

> *G. stercorarius*, Lin. (*putridarius*, Sharp's Cat.)

> **II.** Elytra with nine striae on each between suture and humeral prominence.

Rather more oblong and convex than *G. stercorarius*. Shiny. Colour variable, usually on upper-side blue-black, on under-side steel-blue. Outer-side of mandibles strongly and evenly rounded, indistinctly sinuate at apex ; disc of thorax smooth ; alternate interstices on elytra narrow. Male with two pointed teeth (larger than in *G. putridarius*) beneath posterior femora, and a toothed ridge on hinder side of anterior tibiae. L. 8—11 l. Moderately common.

> *G. mutator*, Marsh.

> **b.** Body nearly hemispherical ; hinder tibiae with two transverse ridges on outer-side.

> **I.** Mesosternum without prominence.

Upper-side blue-black, bluer at margins, under-side blue or violet ; antennae (except first joint) and palpi red-brown. Clypeus with distinct prominence ; thorax diffusely punctured ; elytra with feeble striae, interstices with confused transverse scratches. Spot of hairs on anterior femora golden-yellow. L. 5—8 l. Common. *G. sylvaticus*, Panz.

> **II.** Mesosternum with conical prominence.

> > **1.** Thorax covered with a diffuse punctuation, the interstices being closely and finely punctured.

Blue or blue-black ; antennae and palpi black. Prominence on clypeus feeble ; elytra with very fine, indistinct
striae, interstices either smooth or transversely scratched.
Spot of hairs on anterior femora black. Male with at least
eight teeth on under-side of anterior tibiae. L. 5½—7 l.
Common. *G. vernalis*, Lin.

2. Thorax visibly punctured at sides only.

Similar to *G. vernalis* but much more brilliant, smooth
and shiny, narrower in proportion to length, abdomen impunctate in middle beneath, posterior angles of thorax less
obtuse and rounded. Male with five or six teeth on under-
side of anterior tibiae. L. 5½—7 l. Common.
G. pyrenaeus, Charp.

TROGIDAE.

Club of antennae with plates free and nearly equal in
size. Ligula membranous, hidden by mentum ; lobes of
maxillae horny, inner one armed with a simple hook in
middle and another (usually three-toothed) at apex. Sides
and base of thorax thickly fringed with bristles. Forehead
and thorax uneven. *Trox*, Fab.

Trox.

A. Length 3½—4 lines.

a. Elytra with shallow, feebly punctured striae, alternate interstices somewhat raised and bearing a row of
roundish, smooth elevations, on each of which is a tuft
of longer yellow bristles ; the other interstices granular,
with short bristles.

Dull gray-black. Narrower than *T. sabulosus*, with thorax
rather broader, flatter and rather less closely punctured.
Front of forehead angular in middle. Rare.
T hispidus, Laich.

b. Elytra with broad, shallow, strongly punctured striae,
alternate interstices a little raised and bearing tufts of
very short gray-yellow bristles ; the other interstices
with smaller tufts of similar bristles.

Dull gray-black. Front of forehead rounded. Moderately common. *T. sabulosus*, Lin.

B. Length 2⅔—3 lines.

Dull gray-black. Front of forehead somewhat angular and raised in middle; elytra with shallow, notched striae, interstices alternately, with rows of larger and smaller tufts of very short brownish-yellow bristles. Rather common.

T. scaber, Lin.

MELOLONTHIDAE.

A. Claws unequal.

Mandibles with a broad membranous border on inner-side; labrum not projecting beyond front of clypeus, either horizontal or turned downward. Anterior coxae projecting; posterior coxae not dilated; hinder tibiae with only one or no apical spine. Abdomen apparently with only five ventral segments, the sixth being hidden in fifth; segments soldered together. *Hoplides.*

B. Claws equal.

a. Ventral segments of abdomen not soldered together.

Mandibles with a broad membranous border on inner-side; labrum completely united to clypeus. Anterior coxae projecting; posterior coxae generally dilated; hinder tarsi long and slender. Scutellum triangular. *Sericides.*

b. Ventral segments of abdomen soldered together, with seams effaced in middle.

Mandibles with a very narrow, leathery border to inner-side; labrum prominent, usually deeply emarginate. Anterior coxae scarcely projecting; posterior coxae dilated, but always leaving apex of first abdominal segment free. Scutellum rounded. *Melolonthides.*

HOPLIDES.

Club of antennae three-jointed; all the tibiae without apical spines; posterior tarsi with only one claw.

Hoplia, Ill.

Hoplia.

Head and thorax black; elytra red-brown or brown-red in male, always brown-red in female. Male with scales on upper-side gray, on under-side blue; antennae brown and legs usually black; female with scales green, antennae and legs red. Claw of posterior tarsi split. L. 3½—4 l. Common.

H. philanthus, Sulz.

SERICIDES.

A. Anterior tarsi short.

Antennae with nine joints, club three-jointed.

Homaloplia, Steph.

B. Anterior tarsi long and slender.

Antennae with nine or ten joints, club three-jointed.

Serica, Mc. L.

Homaloplia.

Black, rather shiny; elytra red-yellow, with suture and outer margin black. Hairs on male black on upper-side, gray on under-side; on female all gray. L. $2\frac{1}{2}$—$3\frac{1}{2}$ l. Not very common.   *H. ruricola*, Fab.

Serica.

Oblong, somewhat cylindrical. Brown-red, dull; forehead rather darker. Antennae with nine joints; eyes large and prominent; elytra with striae, interstices narrow and arched. L. 4 l. Moderately common; by sweeping at dusk.   *S. brunnea*, Lin.

MELOLONTHIDES.

A. Third joint of antennae longer than fourth; club in male composed of seven and in female of six joints.

Labial palpi inserted on side margin of labium; tarsal claws with a little pointed tooth at base. *Melolontha*, Fab.

B. Third joint of antennae equal in length to fourth; club three-jointed.

Labial palpi inserted on outer surface of labium near margin; tarsal claws with a little upright tooth at base.

Rhizotrogus, Latr.

Melolontha.

A. Process of last dorsal abdominal segment long, gradually narrowed.

Black; elytra brown-red; abdomen with triangular spots of white hairs at sides. Colour of thorax and elytra slightly variable. L. 12 l. Common.   *M. vulgaris*, Fab.

B. Process of last abdominal segment short, narrowed at base and a little dilated toward apex.

Black ; elytra red-brown, with narrow black outer margin ; abdomen with triangular spots of white hairs at sides. Colour of thorax and elytra rather variable. L. 10—12 l. Not common. *M. hippocastani*, Fab.

Rhizotrogus.

A. Elytra pale yellow, diffusely and feebly punctured ; last dorsal abdominal segment granulate.

Brown; clypeus, sides of thorax and of last upper abdominal segment reddish-yellow. Elytra with four raised longitudinal lines. Forehead, thorax and last upper abdominal segment covered thickly, elytra more diffusely, with long grayish hairs, longer and closer in male than in female. Thorax also with short, whitish-gray pubescence. L. 7—8 l. Common. *R. solstitialis*, Lin.

B. Elytra brown-red or brownish-yellow, rather distinctly punctured ; last dorsal abdominal segment diffusely and finely punctured.

Brown ; clypeus and sides of thorax brown-red ; sides of last upper abdominal segment yellow. Elytra with three raised longitudinal lines and traces of a fourth. Thorax with short, fine gray pubescence, thinly mixed with longer hairs, or (in male) closely covered with long hairs ; elytra covered thinly and last upper abdominal segment more closely with short upright hairs, male with bristles on elytra. L. 5½—7½ l. Rare. *R. ochraceus*, Knoch.

RUTELIDAE.

Antennae with nine joints ; labrum and labium emarginate at apex ; outer-side of mandibles smooth ; elytra with a narrow membranous border at apex. [*Anomalides.*]

A. Posterior femora not thickened or bordered above.

Posterior legs not stronger in proportion than the others. Body rather flat. *Phyllopertha*, Kirby.

B. Posterior femora thickened and bordered above.

Posterior legs strong in proportion to front ones. Body evenly arched. *Anomala*, Leach.

Phyllopertha.

Shiny. Head, thorax and scutellum greenish or bluish-black; elytra reddish-yellow-brown, rarely darker. Body covered with black or gray hairs. L. 4—5 l. Common.

P. horticola, Lin.

Anomala.

Reversed ovate, moderately shiny. Upper-side green, elytra green or yellowish; under-side dark bronze; antennae reddish-yellow, with black club. Inner claw of anterior tarsi gradually thickened toward base; outer claw of front pairs of tarsi split at apex. L. $4\frac{1}{2}$—$6\frac{1}{2}$ l. Common.

A. Frischi, Fab.

CETONIIDAE.

A. Elytra sinuate at side.

Epimera of mesothorax visible between posterior angles of thorax and shoulders of elytra; mesosternum produced into a knob between intermediate coxae; epimera of metathorax and outer margin of dilated posterior coxae visible from above.   *Cetonia*, Fab.

B. Elytra not sinuate at side.

a. Outer lobe of maxillae with long hairs all round; mentum fully as broad as long.

Body smooth. Last joint of palpi oblong, somewhat narrowed toward apex; club of antennae narrow, oblong.

Gnorimus, Serv.

b. Outer lobe of maxillae with long hairs on outer-side and apex; mentum oblong, narrow.

Body hairy. Last joint of palpi long, cylindrical, truncate at apex; club of antennae oval.   *Trichius*, Fab.

Cetonia.

A. Length 7—12 lines.

a. Process of mesosternum almost globular; posterior tibiae with a pointed tooth in middle.

Oblong, flat, very shiny. Upper-side golden-green, sometimes with a coppery reflection; under-side coppery; elytra with small white transverse markings, and with an impression beside suture on hinder part gradually shallower from middle forward; clypeus slightly emarginate in front. L. 7—10 l. Common.   *C. aurata*, Lin.

b. Process of mesosternum dilated, flat, cut off straight in front; posterior tibiae with a rather oblique transverse ridge in middle.

Oblong oval, rather flat, very shiny. Upper-side bronze-green or coppery, under-side dark bronze; elytra with small white transverse markings and with an impression beside suture on hinder part, ceasing rather abruptly in middle; clypeus straight in front. L. 7—12 l. Rather common. *C. floricola,* Herbst.

B. Length 4½—6 lines.

Blackish-bronze, shiny; thorax with two rows of white spots, elytra and last upper abdominal segment with white markings. Upper-side thinly covered with white hairs. Male with a row of white spots along middle of abdomen. Rare. *C. stictica,* Lin.

Gnorimus.

A. Upper-side black.

Black, moderately shiny; elytra and sides of abdomen with whitish spots. Breast and under-side of front pairs of femora covered with gray hairs. L. 8—10 l. Common.
G. variabilis, Lin.

B. Upper-side golden-green.

Shiny. Under-side coppery; elytra and sides of abdomen with white spots, former strongly wrinkled. Breast and under-side of front pairs of femora in female, under-side of body and of all femora in male covered with gray hairs. L. 7—9 l. Scarce. *G. nobilis,* Lin.

Trichius.

A. Abdomen evenly covered with gray hairs.

Black; elytra velvety, with two yellow bands, united at suture. Head, thorax and scutellum closely covered with yellowish, breast with gray hairs. Intermediate tibiae with a long sharp tooth in middle; clypeus rather deeply emarginate in front; elytra half as broad again as thorax, posterior angles of latter rounded. L. 5—6½ l. Not uncommon.
T. fasciatus, Lin.

B. Abdomen thinly covered in middle with gray hairs, bare at sides.

Black; elytra velvety, reddish-yellow, with a spot at shoulder, an abbreviated band in middle and a spot at apex, outer margin and suture black. Head, thorax and scutellum closely covered with brownish-yellow, breast with gray hairs (shorter than in *T. fasciatus*). Intermediate tibiae not toothed; clypeus only slightly emarginate in front; thorax nearly as broad as elytra, with posterior angles almost right angles and with a large spot more feebly covered with hair than the rest and indistinctly punctured on hinder part. Male with posterior femora slightly clubbed toward apex. L. 4½—6 l. Very rare.

T. abdominalis, Men.

STERNOXI.

A. Prothorax more or less closely applied to mesothorax.

a. Eyes oblong oval, more or less large ; labrum distinct.

Insects without power of leaping when placed on back. Antennae inserted in cavities on lowest part of forehead; posterior angles of thorax not produced.   *Buprestidae.*

b. Eyes roundish, small ; labrum obsolete.

Insects sometimes with slight power of leaping when placed on back. Antennae inserted on forehead at inner margin of eyes ; posterior angles of thorax produced.

Eucnemidae.

B. Prothorax not closely applied to mesothorax.

Eyes usually rather large, roundish ; labrum distinct; antennae inserted immediately before eyes ; posterior angles of thorax produced. Insects with strong power of leaping when placed on back.   *Elateridae.*

BUPRESTIDAE.

A. Tarsi long.

a. Scutellum triangular.

Mentum rather broader than long, apex truncate or nearly so. Hinder process of prosternum flat, broad, dilated on each side behind anterior coxae, apex broad, angular. First and second abdominal segments soldered together but with a visible seam. Base of thorax straight ;

first tarsal joint somewhat elongate.   *Anthaxia*, Esch.

b. Scutellum very broad at base, suddenly and sharply pointed at apex.

Mentum transverse, obtusely triangular. Prosternum with a process in front, partly covering mouth and separated by a furrow ; its hinder process flat, moderately broad, scarcely dilated before the rather broad, slightly rounded apex. · First and second abdominal segments soldered together, seam scarcely visible. Base of thorax sinuate on each side ; first joint of posterior tarsi very elongate.

Agrilus, Sol.

B. Tarsi short.

a. Thorax as broad as, or somewhat broader than long ; body elongate.

Head horizontal ; antennae received in grooves, last four joints saw-like ; all coxae placed rather far apart ; femora dilated inward, receiving tibiae ; first and second abdominal segments soldered together, seam distincts.

Aphanisticus, Latr.

b. Thorax much broader than long ; body broad.

Head vertical ; antennae received in traces of grooves, last five joints sawlike ; hinder pairs of coxae placed much farther apart than anterior pair ; femora only slightly dilated, scarcely receiving tibiae ; first and second abdominal segments soldered together, seam scarcely visible.

Trachys, Fab.

Anthaxia.

Rather elongate. Golden-green ; head, thorax and part of abdomen sometimes purplish. Closely punctured in wrinkles ; thorax almost double as broad as long, apex sinuate on each side, disc with a shallow central furrow and an impression on each side, base narrowed more than apex. L. 2—2¼ l.   Rare.   *A. nitidula*, Lin.

Agrilus.

A. Claws split at apex.

a. Elytra with a white spot of scalelike hairs on each.

Green or blue ; each side of abdomen with three white spots. Apex of elytra rounded and finely notched. L. 4— 5½ l.   Not common.   *A. biguttatus*, Fab.

b. Elytra with an oblong spot (pointed behind and frequently absent) of gray hairs before apex.

Upper-side purple, dull, with coppery reflection ; underside bronze-green. Thorax transversely wrinkled, uneven, with a fine, raised curved line in posterior angles ; elytra closely granulate ; front of prosternum deeply emarginate ; last ventral segment of abdomen rounded.　L. 3½—4 l. Scarce.　　　　　　　　　　　　　　　*A. sinuatus*, Ol.

B. Claws with a broad and more or less blunt tooth at base.

　a. Last ventral segment of abdomen rounded, without impression.

Olive-green, bluish or bronze ; forehead bluish or. coppery ; under-side with brassy reflection.　Thorax much broader than long, coarsely wrinkled irregularly, with a small raised ridge in right angled posterior angles ; elytra rather strongly wrinkled, at apex strongly rounded and slightly divergent, finely but distinctly toothed.　L. 2½— 3½ l.　Rare.　　　　　　　　　　　　　*A. viridis*, Lin.

　b. Last ventral segment of abdomen emarginate, with an oblong impression in male.

　　I. Antennae sawlike (in male nearly comblike) in middle ; first abdominal segment in male with two small prominences in middle of apex.

Bluish, green or bronze.　Thorax broader than long, coarsely wrinkled and distinctly punctured, with a ridge (curved in front only and reaching nearly to middle) in posterior angles ; elytra moderately closely and strongly granulate, at apex truncate and very indistinctly toothed. L. 2—2½ l.　Common.　　　　　*A. angustulus*, Ill.

　　II. Antennae strongly dilated in middle, fourth and following joints broad triangular ; first abdominal segment in male without prominences.

Narrow, gradually narrowed behind.　Olive-green ; disc of thorax blackish in front.　Thorax much broader than long, coarsely and not very closely wrinkled, with a more or less distinct, curved ridge in posterior angles ; elytra granulate, apex rounded and extremely finely toothed.　L. 2½ l. Moderately common.　　　　　　　　*A. laticornis*, Ill.

Aphanisticus.

Oblong oval. Black, with slight bronze reflection. Thorax transverse, with three transverse impressions (the two hinder ones somewhat indistinct), posterior angles sharp right angles. L. 1—1¼ l. Moderately common.

A. pusillus, Ol.

Trachys.

A. Length 1¼—1½ lines.

a. Head and thorax black, metallic.

Elytra black, with bluish reflection, with four waved bands of whitish hairs, uneven, shoulders raised, punctuation indistinct and diffuse. L. 1½ l. Moderately common.

T. minutus, Lin.

b. Head and thorax dark coppery.

Elytra dark blue, without hair, strongly but not deeply punctured in rows. Prosternum rather broad, nearly parallel-sided. L. 1¼ l. Rare. *T. troglodytes*, Gyll.

B. Length ⅘ line.

Black, rather metallic. Elytra triangular, moderately strongly punctured in irregular rows, sides with a more or less distinct raised line. Not uncommon. *T. nana*, Fab.

EUCNEMIDAE.

A. Labrum present.

Last joint of palpi hatchet-shaped ; antennae short, first joint large, middle joints small, roundish, last three joints forming a club, under-side of thorax with angular grooves for their reception ; prosternum flat, oblong, tolerably parallel-sided, truncate in front, rounded behind, with sloping sides, bordered behind and at sides by an impressed line ; tarsi long, first joint as long as next three together, third slightly, fourth distinctly bilobed. *Throscus*, Latr.

B. Labrum absent.

a. Last joint of maxillary palpi oval.

Antennae comblike in both sexes, but more so in male ; under-side of thorax without any grooves for reception of antennae ; tarsi broad, compressed. *Melasis*, Ol.

b. Last joint of maxillary palpi hatchet-shaped.

I. No grooves for reception of antennae beneath thorax ; posterior coxae broad and flat, but not dilated over femora.

Antennae comblike in male, sawlike in female ; first tarsal joint long, fourth short heart-shaped ; claws toothed, comblike. *Cerophytum*, Latr.

II. Under-side of thorax with grooves for reception of antennae ; posterior coxae dilated over femora.

Antennae sawlike or comblike ; tarsi with first joint long, fourth heartshaped. *Microrhagus*, Esch.

Throscus.

A. Thorax rather diffusely and indistinctly punctured ; length 1¼—1½ lines.

a. Eyes divided until middle.

Reddish-brown, with fine, silky yellowish-gray pubescence. Head with two parallel, slightly raised lines; elytra with fine punctured striae, interstices finely shagreened and diffusely punctured. Common. *T. dermestoides*, Lin.

b. Eyes divided until considerably beyond middle.

Similar to *T. dermestoides*, but with the two frontal ridges more distinct and extended backward to thorax, elytra more pointed behind, with striae feebler and punctures of interstices rather clearer, the surface being not so coarsely granulated. L. 1¼ l. Rare. *T. carinifrons*, Bonv.

B. Thorax distinctly and somewhat closely punctured ; length ¾—1 line.

a. Head without raised lines.

Reddish-brown. More oval than *T. dermestoides*, more evenly arched ; pubescence finer and whitish-gray ; elytra only indistinctly shagreened, with very fine punctured striae, except near base and suture, where are only rows of punctures, interstices finely but distinctly punctured. L. ¾—1 l. Rare. *T. obtusus*, Curt.

b. Head with two slightly raised lines.

Oblong. Reddish-brown, pubescence rather short and close. Thorax rather short, sides dilated before acute posterior angles; elytra oblong, truncate at apex, with punctured striae, feebler toward suture, interstices indistinctly and closely punctured. L. 1 l. Rare.

T. elateroides, Heer.

Melasis.

Pitch-black; antennae brown. Punctuation rough; pubescence scanty, gray; elytra with striae. L. 3—4 l. Not common.

M. buprestoides, Lin.

Cerophytum.

Cylindrical. Black; mouth, antennae and legs reddish. Punctuation close and strong; pubescence fine; elytra with punctured striae, interstices wrinkled. L. 3—3½ l. Rare.

C. elateroides, Latr.

Microrhagus.

Almost cylindrical. Black; tibiae and tarsi reddish. Scantily covered with gray hairs; punctuation deep and wrinkled. Thorax with two small depressions on disc; elytra without distinct striae, except beside suture. Antennae of male comblike. L. 1½—2 l. Rare.

M. pygmaeus, Fab.

ELATERIDAE.

A. Under-side of thorax with deep grooves for reception of antennae.

Antennae with first joint large, cylindrical, second and third small, globular, fourth much broader and double as long as third; antennal grooves not quite reaching anterior coxae; tarsi laterally compressed, simple. *Lacon*, Germ.

B. Under-side of thorax with feeble or no grooves for reception of antennae.

a. Posterior coxae dilated, suddenly narrowed outward.

I. Under-side of thorax with traces of grooves for reception of antennae.

Antennae with second joint small, third not much longer, conical, rarely triangular, rest sawlike; head bent downward, forehead projecting over labrum; tarsi simple.

Elater, Lin.

II. Under-side of thorax with no traces of grooves for reception of antennae.

1. Scutellum oval or roundish quadrangular ; sternal spine moderately long, thin.

A A. First joint of antennae only moderately large.

Head bent downward, forehead projecting over labrum ; seams beneath thorax double; tarsi elongate, first nearly as long as next two together.   *Megapenthes*, Kies.

B B. First joint of antennae large.

Head only slightly bent downward, forehead raised above labrum; tarsi bristly, first joint somewhat longer than second.   *Cryptohypnus*, Germ.

2. Scutellum heartshaped ; sternal spine short and thick.

Third joint of antennae rather longer than second and about as long as fourth; head somewhat bent downward, forehead projecting over labrum.   *Cardiophorus*, Esch.

b. Posterior coxae gradually narrowed outward.

I. Head considerably sunk in thorax ; eyes not strongly convex.

1. Last joint of maxillary palpi more or less hatchet-shaped.

A A. Third tarsal joint without large membranous appendage.

a a. Tarsal claws comblike.

Second and third joints of antennae small ; forehead projecting above labrum ; seams beneath thorax double in front ; tarsi strong, bristly, simple.   *Melanotus*, Esch.

b b. Tarsal claws simple.

A a. Lower part of forehead and labrum only moderately deflexed, mouth in front of head

A 1. Under-side of thorax with distinct traces of grooves for reception of antennae.

Tarsi strong, simple.   *Limonius*, Esch.

B 2. Under-side of thorax without any distinct traces of grooves for reception of antennae.

a 1. Forehead with a distinct transverse ridge.

Tarsi strong, usually somewhat dilated. *Athous,* Esch.

b 1. Forehead not ridged.

A 2. Posterior coxae not much dilated and simple.

Tarsi rather long and thin, rather bristly.
Corymbites, Latr.

B 2. Posterior coxae rather broad, toothed behind.

Tarsi rather long, bristly. *Ludius,* Latr.

B b. Lower part of forehead and labrum much deflexed, mouth on or nearly on under surface of head.

A 1. Sides of thorax blunt, marginal line bent on to under surface in front.

Antennae sawlike; seams beneath thorax double, forming traces of antennal grooves in front. *Agriotes,* Esch.

B 1. Sides of thorax sharp, bordered.

a 1. Posterior coxae scarcely narrowed outward.

Antennae strongly sawlike, first joint moderately long, only slightly thickened, second and third small, about equal; seams beneath thorax double, forming traces of antennal grooves in front. *Sericosomus,* Redt.

b 1. Posterior coxae considerably narrowed outward.

Antennae almost threadlike, second joint oblong, rather longer than third; seams beneath thorax double, interstice not deep in front. *Dolopius,* Esch.

B B. Third tarsal joint with a large membranous appendage beneath, covering under-side of the very small fourth joint.

Labrum and mouth inflexed on to lower surface of head; under-side of thorax with traces of grooves for reception of antennae; tarsal claws comblike on inner-side.
Ctenonychus, Steph.

2. Last joint of maxillary palpi ovate, pointed.

Head bent downward; under-side of thorax without antennal grooves; fourth tarsal joint dilated or simple, claws sawlike.   *Adrastus*, Esch.

II. Head prominent; eyes strongly convex.

Anterior margin of forehead sharp, raised; seams beneath thorax simple; intermediate coxae much approximated.
Campylus, Fisch.

Lacon.

Pitch-black, entirely covered with thick gray and brown marbled pubescence. Elytra with feeble punctured striae, interstices finely punctured.   L. 5—7 l.   Common.
L. murinus, Lin.

Elater.

A. Elytra bright red, either unicolorous, with a common black spot, or with apex black.

a. Thorax short.

I. Central furrow on thorax distinct throughout.

Black, elytra scarlet. Hairs black, on under-side very fine. Thorax closely and strongly punctured. L. 5—6 l. Rare.   *E. sanguineus*, Lin.

II. Central furrow on thorax distinct at base only.

1. Apex of elytra red.

A A. Upper-side covered with pale hairs.

Black, elytra scarlet. Hairs on under-side close-lying and silky. Thorax closely and strongly punctured. Body rather broader than in *E. sanguineus*, thorax somewhat more convex, striae on elytra not quite reaching base. L. 5—6 l. Rather common.   *E. lythropterus*, Germ.

B B. Upper-side covered with a mixture of light and dark hairs.

Black; elytra scarlet, generally with a common, black, long oval spot. Hairs on head and thorax black and gray mixed, on elytra gray or brownish. Thorax closely and strongly punctured, a little more scantily on hinder part of disc, without central furrow; elytra with strong punctured striae. L. 4—5½ l. Not common.
E. sanguinolentus, Schr.

2. Apex of elytra black.

Rather narrow. Black, with black hairs ; elytra bright red, with apex more or less black ; tarsi pitchy. Thorax rather less convex than in *E. sanguinolentus*, more finely punctured, and more straightly narrowed toward front ; joints of antennae longer, tarsi thinner and elytra rather flatter. L. 4½—5½ l. Scarce. *E. pomonae*, Steph.

b. Thorax longer.

Nearly parallel-sided, flat. Black, with brown hairs ; elytra bright red ; antennae and legs pitch-black, second and third joints of former and the tarsi reddish. Thorax long, sides almost parallel from base until beyond middle, thence gradually narrowed forward, very closely punctured throughout, disc rather shiny, sides dull, with a central furrow behind and a depression above scutellum ; elytra with feeble punctured striae, interstices flat. L. 5½ l. Rare. *E. coccinatus*, Rye.

B. Elytra unicolorous brownish-red.

Black ; elytra brownish-red. Hairs brown. L. 4—5 l. Moderately common. *E. pomorum*, Herbst.

C. Elytra brownish-red, with apex black.

a. Elytra black at apex only.

Head and thorax black, latter rather more diffusely punctured in middle than at sides, without central furrow, covered with black hairs. L. 3½ l. Scarce.
E. elongatulus, Ol.

b. Elytra with last third part black.

Head and thorax black, latter rather closely and finely punctured, with a feeble central furrow, hairs gray. L. 3½—4 l. Rather common. *E. balteatus*, Lin.

D. Elytra black, with the base, a longitudinal spot near scutellum and outer margin brownish-yellow.

Head and thorax black, latter rather diffusely and moderately finely punctured, with a shallow central furrow, hairs black. L. 3½ l. Rare. *E. tristis*, Lin.

E. Elytra entirely black.

a. Thorax broader than long ; antennae longer than head and thorax ; third joint about double as long as second.

Black, shiny. Thorax diffusely and moderately finely

D 2

punctured; interstices on elytra usually rather strongly wrinkled transversely. L. 2¾—3½ l. Not uncommon.

E. nigrinus, Herbst.

b. Thorax scarcely broader than long; antennae as long as head and thorax, third joint scarcely longer than second.

Black, shiny. Thorax somewhat closely and strongly punctured; interstices on elytra diffusely punctured, only slightly wrinkled. L. 4½ l. Rare.   *E. aethiops*, Lac.

Megapenthes.

A. Antennae strongly sawlike, third joint double as long as second.

Black; thorax (with reflexed margins) red, rather diffusely punctured. L. 4½ l. Not common.

M. sanguinicollis, Panz.

B. Antennae feebly sawlike, third joint small.

a. Body gradually narrowed from front backward, dull.

Black; tibiae pitchy. Thorax closely and shallowly punctured. L. 4—5 l. Very rare.   *M. lugens*, Redt.

b. Body parallel-sided, moderately shiny.

Black; tibiae reddish. Forehead with an indistinct longitudinal prominence; thorax moderately closely and strongly punctured. L. 3—4 l. Rare. *M. tibialis*, Lac.

Cryptohypnus.

A. Length 2¼—2¾ lines.

a. First joint of antennae shorter than third joint.

Black, with slight bronze reflection. Pubescence close, fine, gray; punctuation very close and fine. Thorax as long as broad; elytra with deep punctured striae; first joint of antennae broader than long, shorter than third joint; prosternal process covering only base of mouth parts. Scarce.

C. maritimus, Curt.

b. First joint of antennae longer than third joint.

Black, with bronze reflection; base of antennae and legs red. Pubescence golden-yellow. Thorax broader than long, punctuation diffuse, moderately fine but deep; elytra with impunctate striae, interstices diffusely and very finely punctured; first joint of antennae longer than broad and longer than third joint; prosternal process entirely covering mouth parts. Common. *C. riparius*, Fab.

B. Length $1\frac{1}{4}$—2 lines.

   a. Elytra black, with yellow spots.

      I. Prosternal process entirely covering mouth parts.

         1. Femora brown in middle.

Elongate, only slightly convex. Black; elytra with variable yellow spots at base and toward apex; base of antennae and legs yellow, femora brown in middle. Very similar to *C. pulchellus* but with thorax longer, its posterior angles shorter and not at all divergent, ridge not reaching middle, disc more strongly punctured in wrinkles; elytra with striae distinct at apex, interstices ridged until beyond middle. L. $1\frac{3}{4}$—2 l. Rare. *C. sabulicola*, Boh.

         2. Femora entirely yellow.

Black, dull; elytra with several variable yellow spots at base and toward apex; base of antennae and legs (with anterior coxae) yellow. Thorax longer than broad, with a slightly raised central line, very closely and moderately strongly punctured, posterior angles slightly divergent, ridge reaching middle; elytra with striae rather indistinct at apex, interstices ridged to middle. L. $1\frac{1}{4}$—$1\frac{2}{3}$ l. Rare. *C. pulchellus*, Lin.

      II. Prosternal process covering base of mouth parts only.

Black; elytra with two yellow spots on each. Thorax as broad as long, without raised central line, finely punctured; first joint of antennae shorter than third joint. L. $1\frac{1}{2}$ l. Common. *C. quadripustulatus*, Fab.

   b. Elytra unicolorous black.

Black; first joint of antennae black; femora darker in middle; hairs gray. Thorax very finely, closely wrinkled, ridges from posterior angles prolonged beyond middle. L. $1\frac{1}{2}$ l. Common. *C. dermestoides*, Herbst.

Cardiophorus.

A. Thorax entirely red.

Head and elytra black.   L. 3½ l.   Very rare.
                                 C. thoracicus, Fab.

B. Thorax red, base and front third part black.

Head and elytra black.   L. 2¾ l.   Very rare.
                                 C. ruficollis, Lin.

C. Thorax entirely black.

Gray-black, with close grayish pubescence.   L. 3½—4 l.
Not very common.   *C. asellus*, Er.

Melanotus.

A. Scutellum as broad as long.

Black. Thorax strongly and very closely punctured all
over, with a raised central line.   L. 5½—6½ l.   Not common.   *M. punctolineatus*, Pel.

B. Scutellum much longer than broad.

   a. Sides of thorax scarcely rounded.

Pitch-black or pitch-brown.   Thorax punctured rather
closely at sides, more diffusely in middle; elytra 3½ times
as long as thorax.   Antennae of male with long hairs.   L.
6½—9 lines.   Rare.   *M. castanipes*, Payk.

   b. Sides of thorax evenly rounded.

Similar to *M. castanipes*, but with elytra only three times
as long as thorax and punctuation feebler.   L. 5—8 l.
Common.   *M. rufipes*, Herbst.

Limonius.

A. Interstices of punctures on thorax not larger than punctures themselves.

Black, with some bronze reflection ; hairs gray.   Sternal
spine furrowed.   L. 4—5 l.   Rather common.
                                 L. cylindricus, Payk.

B. Interstices of punctures on thorax much larger than
punctures themselves.

   a. Tarsal claws somewhat dilated until near middle,
then suddenly narrowed.

Black, with very little, if any, bronze reflection ; tibiae
and tarsi dark brown ; pubescence scanty, gray.   Thorax

diffusely and strongly punctured, with scattered black hairs; sides of elytra straight. Antennae of male strongly saw-like. L. 2½ l. Common. *L. minutus,* Lin.

> *b.* Tarsal claws long and fine, with traces of a tooth at base only.

Black, with greenish-bronze reflection; tibiae and tarsi yellow; pubescence thick, yellowish. Thorax moderately closely and finely punctured, without scattered black hairs; sides of elytra slightly curved. L. 3—3½ l. Rare.
L. parvulus, Panz.

Athous.

A. Thorax without central furrow.

> *a.* Fourth tarsal joint considerably shorter and narrower than third joint.

> > *I.* Antennae strongly sawlike from third joint.

> > > *1.* Under-side brown or rust-red; third joint of antennae fully as long as fourth.

Brown or rust-red; head and thorax rather darker. Thorax and elytra rather closely covered with long gray hairs, latter with two more or less bare oblique bands enclosing a rhombic spot of stronger pubescence; thorax more closely punctured at sides than on disc, posterior angles moderately sharp, feebly ridged. L. 9 l. Rare.
A. rhombeus, Ol.

> > > *2.* Under-side black; third joint of antennae shorter than fourth.

Black, more or less shiny, elytra sometimes yellow-brown; pubescence rather close, long, gray. Thorax moderately finely and closely punctured, posterior angles rather sharp, with a somewhat sharp ridge; second and third tarsal joints heart-shaped. Female broad. L. 4½—7 l. Rather common. *A. niger,* Lin.

> > *II.* Antennae feebly sawlike, third joint reversed conical, without sharp inner angle.

> > > *1.* Third joint of antennae much longer than second.

Pitch-brown or black; elytra lighter brown; abdomen either entirely or with margins and sides of segments and its apex reddish; tibiae and tarsi sometimes reddish. Thorax closely and rather strongly punctured, posterior angles nearly right angles, not ridged. L. 5—6½ l. Common.   *A. haemorrhoidalis*, Fab.

2. Third joint of antennae only slightly, or not at all longer than second.

Very similar to *A. haemorrhoidalis*, but rather broader. Black or dark brown; margins of thorax, elytra (except margins), under-side (except metasternum and base of abdomen), and antennae often brownish-yellow; legs always lighter or darker brownish-yellow, femora at most slightly darker. L. 4—5½ l. Rather common.

A. vittatus, Fab.

b. Second to fourth tarsal joints gradually diminishing in length, all distinctly dilated.

I. Third joint of antennae scarcely longer than second.

Lighter or darker brownish-yellow; head, disc of thorax, breast and base of abdomen blackish or dark brown. Forehead scarcely impressed; thorax diffusely punctured on disc, rather more closely at sides; posterior angles nearly right angles, not ridged; antennae feebly sawlike. L. 3½—4 l. Rare.   *A. subfuscus*, Müll.

II. Third joint of antennae distinctly longer than second.

Brown-red, only slightly shiny; elytra with suture and outer margin narrowly, and a streak from shoulder to apex yellowish-red. Forehead strongly impressed; thorax longer than broad, strongly and not very closely punctured, posterior angles not ridged; elytra with punctured striae; antennae feebly sawlike. L. 4½ l. Rare.

A. difformis, Lac.

B. Thorax with a more or less distinct central furrow and a depression on each side at base.

a. Antennae feebly sawlike, third joint reversed conical, without sharp inner angle; fourth tarsal joint very much smaller than third.

Dull. Head and thorax black, sides of latter lighter; elytra yellow or yellow-brown, with outer margin (male) or outer margin and inner-side (female) blackish; under-side black, abdomen (especially margins of segments), tibiae and tarsi more or less brown. Forehead impressed; central furrow on thorax abbreviated and feeble, punctuation very close and rather strong; third joint of antennae more than double as long as second. Male elongate, flattish, with sides of thorax straight and of elytra parallel; female moderately elongate, convex, with sides of thorax slightly rounded and elytra sometimes very slightly dilated beyond middle. L. 4—4½ l. Common. *A. longicollis*, Ol.

b. Antennae strongly sawlike from third joint; second to fourth tarsal joints gradually diminishing in length and breadth.

Black, dull; thorax and elytra sometimes brown. Pubescence on head and thorax only moderately close but long; elytra partly bare but with base, a curved spot from shoulder to suture, a zigzag band behind middle and the apex closely covered with gray hairs; third joint of antennae fully double as long as second. L. 6—7 l. Very rare. *A. undulatus*, De G.

Corymbites.

A. Third joint of antennae almost as broad as fourth.

a. Second joint of antennae roundish.

I. Antennae longer than head and thorax, comblike in male; sawlike in female.

1. Thorax with strong central furrow.

A A. Male with comblike processes on antennae fully double as long as the joints; female with interstices on elytra convex.

Brassy, with some greenish reflection. Joints of antennae in female cut off obliquely at apex, inner apical angle prominent. L. 6—8 l. Not uncommon.
C. pectinicornis, Lin.

B B. Male with comblike processes on antennae scarcely longer than the joints; female with interstices on elytra flat.

Coppery, violet or greenish-bronze; elytra yellowish from base to beyond middle. Joints of antennae in female cut off straight, inner apical angle moderately sharp. L. 6—7 l. Rather common. *C. cupreus*, Fab.

The variety *C. aeruginosus*, Germ., has whole of elytra coloured like thorax.

2. Thorax without central furrow.

Black; elytra brownish-yellow, apex black; head and thorax covered with yellow pubescence. L. 4—4½ l. Rare.
                                        C. castaneus, Lin.

II. Antennae as long as head and thorax, strongly sawlike in both sexes.

Metallic-brown, with violet or bronze reflection; pubescence rather strong, closelying, gray, usually clouded in parts. L. 6—7 l. Scarce.   *C. tessellatus*, Lin.

b. Second joint of antennae conical.

Narrow, parallel-sided. Black; somewhat leaden; elytra and legs sometimes lighter; pubescence short. Antennae feebly sawlike. Head closely and strongly punctured; thorax much longer than broad. L. 3½—4 l. Rather common.   *C. quercus*, Gyll.

The variety *C. ochropterus*, Steph., has the elytra pale red-yellow or ochre-yellow, legs pitch-black, with tibiae paler and antennae black.

B. Third joint of antennae much narrower than fourth.

a. Seams beneath thorax double; upper-side thickly covered with cloudy, silky pubescence.

Dark brown; legs lighter. Thorax about as long as broad, posterior angles not projecting outward. L. 4½ l. Common.   *C. holosericeus*, Fab.

b. Seams beneath thorax simple; upper-side with scanty or no pubescence.

Thorax about as long as broad, posterior angles projecting outward; antennae feebly sawlike.

I. Elytra black, bluish or bronze.

1. Upper-side without pubescence.

Rather broad. Black, violet, bluish or bronze; legs sometimes red, interstices on elytra flat. L. 5—7 l. Rather common.   *C. aeneous*, Lin.

2. Upper-side with scanty pubescence.

A A. Thorax rather shiny, without central furrow.

Brown, with greenish-bronze reflection; legs red; pubescence rather coarse, very short, golden-yellow. Sides of thorax slightly rounded; elytra scarcely broader than thorax, interstices flat. L. 5 l. Not common.
                                        C. metallicus, Payk.

B B. Thorax dull, with a shallow central furrow.

Blackish, with bronze reflection ; legs sometimes lighter ; pubescence irregular, silky, gray. Sides of thorax very slightly rounded ; elytra broader than thorax, interstices moderately arched. L. 5½—6½ l. Not common.

C. impressus, Fab.

II. Elytra black, with a round red spot on shoulder, or sometimes entirely red-yellow.

Head and thorax black ; legs lighter. Sides of thorax rounded, punctuation diffuse. L. 3—3½ l. Not common.

C. bipustulatus, Lin.

Ludius.

Broad. Black ; thorax (except base), and elytra yellow-red, former sometimes entirely black ; pubescence rather close, yellow. L. 7—8 l. Very rare. *L. ferrugineus*, Lin.

Agriotes.

A. Second joint of antennae much shorter than fourth.

Elongate. Brown; pubescence closelying, gray. Thorax closely and strongly punctured ; with a central furrow behind, posterior angles ridged ; elytra as broad as thorax, slightly dilated in middle. L. 6—7 l. (? Brit.)

A. pilosus, Panz.

B. Second joint of antennae as long, or nearly as long as fourth.

a. Posterior coxae only slightly dilated on inner part, their inner third part less than twice as broad as their outer third part. •

I. Posterior angles of thorax distinctly ridged.

1. Elytra light brown, with a dark shade on disc ; length 2½—3 lines.

Oblong. Head and thorax black, base and apex of latter yellowish-brown. Pubescence rather thick, gray. Thorax longer than broad, rather closely punctured, with an indistinct central furrow at base, sides nearly straight, posterior angles directed backward ; elytra long. Common.

A. sputator, Lin.

2. Elytra yellowish, alternate interstices brown ; length 4 lines.

Oblong. Head and thorax dark brown, latter with margins lighter. Pubescence gray. Thorax nearly as long as broad, closely punctured, sides slightly rounded, posterior angles projecting somewhat outward. Common.

A. *lineatus*, Lin.

II. Posterior angles of thorax not or only indistinctly ridged.

Short, strongly convex. Dark brown ; elytra lighter ; pubescence gray. Thorax dull, much broader than long, very closely and rather strongly punctured with a more or less distinct central furrow toward base, posterior angles projecting outward. L. 4—4½ l. Common.

A. *obscurus*, Lin.

b. Posterior coxae dilated on inner part more or less suddenly, their inner third part twice as broad as their outer third part.

I. Posterior angles of thorax finely ridged.

1. Thorax strongly punctured, unicolorous black.

Oblong, parallel-sided, not very convex, rather dull. Black, with gray pubescence ; antennae, palpi and legs rust-red, femora brownish. Mandibles compressed ; thorax as broad as long, closely and strongly punctured, sides straight, parallel ; elytra rather flat on disc, with punctured striae, interstices granulate. L. 4—4¾ l. Rare.

A. *sordidus*, Ill.

2. Thorax finely punctured, with posterior angles rust-red.

Elongate, not very convex. Dark brown or black ; posterior angles of thorax and the elytra brownish-yellow, latter blackish at base, suture and side margin ; pubescence scanty. Antennae rather strongly sawlike ; thorax longer than broad, diffusely and rather strongly punctured, side marginal line bent on to under side in front and interrupted in middle ; elytra considerably broader than thorax. L. 2¾ —3 l. Rather common.   A. *sobrinus*, Kies.

II. Posterior angles of thorax not ridged.

Elongate, convex. Dark brown or black ; elytra either dark brown or brownish-yellow ; pubescence scanty. Antennae scarcely sawlike ; thorax longer than broad, diffusely and rather strongly punctured, side marginal line interrupted in middle; elytra scarcely broader in front than base of thorax. L. 1¾—2 l. Rather common.

A. pallidulus, Ill.

Sericosomus.

Oblong, pubescence gray; thorax rather longer than broad, closely punctured, with a more or less distinct central furrow. Male much narrower than female, black, thorax usually with some greenish reflection, elytra brownish-red or dark brown, suture generally blackish; female brownish rust-red, usually with head, a more or less extensive longitudinal spot on middle of thorax and another at basal margin, scutellum, hinder part of breast and base of abdomen black. L. 3¼—4½ l. Moderately common.

S. brunneus, Lin.

Dolopius.

Elongate, rather flat. Dark brown; margins of thorax paler; elytra yellowish-brown, suture (more widely toward base) and middle of outer margin dark; base of antennae and legs yellowish; hairs gray. L. 3 l. Common.

D. marginatus, Lin.

Ctenonychus.

Elongate, nearly cylindrical. Black or pitch-black, closely covered with strong, gray pubescence; antennae and tarsi reddish. Thorax convex, longer than broad; elytra with rows of fine punctures. L. 4½—5 l. Scarce.

C. filiformis, Fab.

Adrastus.

Black; elytra brownish-yellow, with suture, outer margin and hinder part (to a variable extent) blackish; base of antennae and legs reddish-yellow, femora brown; pubescence gray. Third joint of antennae double as long as second. L. 2—2½ l. Moderately common. *A. limbatus*, Fab.

Campylus.

Black: front of head and the thorax (wholly or at margins) red; elytra of male yellow, with or without black suture, of female black, with yellow border. Thorax with a central furrow and a deep oblique impression on each side behind. L. 4½—5½ l. Rather common.

C. linearis, Lin.

MALACODERMA.

A. Prosternum produced into a distinct point between coxae.

a. Labial palpi not very strongly developed.

I. Socket holes for anterior coxae widely open behind and within, almost effaced.

1. Abdomen with five ventral segments.

A A. Maxillary palpi distinctly four-jointed, apex without spines.

a a. Mandibles prominent.

Maxillae usually pointed or split at apex; ligula split; first four tarsal joints with large membranous soles.

Dascillidae.

b b. Mandibles hidden.

Lobes of maxillae short, narrow, roundish and fringed at apex; ligula truncate; labial palpi often forked; first four tarsal joints simple, fourth usually bilobed. *Cyphonidae.*

B B. Maxillary palpi apparently three-jointed, with three short blunt spines at apex.

Mandibles hidden; lobes of maxillae small, inner one indistinct; ligula split into four at apex; tarsi slender, joints simple, first one elongate. *Eubriidae.*

2. Abdomen with seven (rarely six) ventral segments.

A A. Antennae inserted close together; mandibles small.

a a. Eyes moderately large.

Femora inserted at, or close to apex of trochanter. Head more or less hidden under thorax; female with elytra and wings. Body not luminous. *Lycidae.*

b b. Eyes very large.

Femora inserted on outer-side of trochanters. Thorax generally much dilated, covering head; female usually without elytra and wings. Body generally with power of emitting light. *Lampyridae.*

B B. Antennae inserted somewhat apart; mandibles large.

Femora inserted on outer-side of trochanters.

a a. Labrum distinct.

Antennae strong, often sawlike or comblike; head sunk to eyes in thorax. Female sometimes without elytra or wings. *Drilidae.*

b b. Labrum obsolete.

Antennae threadlike; head more or less prominent; female with elytra and wings. *Telephoridae.*

II. Socket holes for anterior coxae large and widely open behind, but the coxae somewhat enclosed by anterior margin of mesosternum.

Antennae almost always inserted near margin of forehead beneath eyes; clypeus and labrum distinctly separate; epimera of mesothorax distinct; abdomen with either six or five ventral segments; fourth tarsal joint usually small, not dilated. *Melyridae.*

b. Labial palpi very strongly developed.

Antennae inserted on sides of forehead before eyes, often thickened toward apex; clypeus and labrum distinct; episterna and epimera of mesothorax distinct; socket holes for anterior coxae open behind; abdomen with six or five ventral segments; tarsi dilated, with fleshy soles beneath, fourth joint sometimes very small. *Cleridae.*

B. Prosternum not produced into a distinct point behind.

Socket holes for anterior coxae almost effaced; abdomen with seven or six ventral segments; apical spines of tibiae indistinct; tarsi slender. *Lymexylonidae.*

DASCILLIDAE.

Mandibles hollowed out on inner-side, upper margin toothed before apex; outer lobe of maxillae split; ligula split into four parts. *Dascillus,* Latr.

Dascillus.

Oblong-ovate, convex. Brown, covered with very thick, fine, gray pubescence; apex of abdomen, and in female antennae, legs and elytra yellow-brown. L. 4—5½ l. Common. *D. cervinus,* Lin.

CYPHONIDAE.

A. Posterior femora only moderately thickened; posterior tibiae with feeble apical spines.

a. Mandibles curved, sharp at apex.

I. Joints of antennae long, cylindrical.

1. First joint of posterior tarsi strongly elongate, second produced at inner angle.

Second and third joints of antennae small, about equal in length ; labial palpi forked. Thorax twice or rather more than twice as broad as long, anterior margin rounded.

Helodes, Latr.

2. First joint of posterior tarsi moderately long, second simple.

Third joint of antennae longer than second ; labial palpi forked, third joint not thickened toward apex. Thorax rather more than twice as broad as long, anterior margin flatly rounded.　*Microcara,* Th.

II. Joints of antennae reversed conical.

Third joint of antennae longer than second ; labial palpi simple ; first tarsal joint somewhat elongate. Thorax three or four times as broad as long, anterior margin sinuate on each side.　*Cyphon,* Payk.

III. Joints of antennae saw-like.

Second and third joints of antennae very small, third shorter than second ; labial palpi forked ; first tarsal joint elongate. Thorax three times as broad as long, anterior margin sinuate on each side.　*Prionocyphon,* Redt.

b. Mandibles triangular, blunt at apex.

Joints of antennae cylindrical, first two large and thick, third much thinner and shorter than second ; labial palpi simple ; first tarsal joint elongate. Thorax three times as broad as long, anterior margin sinuate on each side.

Hydrocyphon, Redt.

B. Posterior femora much dilated ; posterior tibiae with one of the apical spines very long.

Joints of antennae cylindrical, third longer than second ; mandibles triangular, blunt at apex ; labial palpi forked ; first tarsal joint strongly elongate. Thorax three times as broad as long, anterior margin sinuate on each side.

Scirtes, Ill.

Helodes.

A. Head, thorax and scutellum red-yellow.

Oblong-ovate. Red-yellow. Eyes, apex and suture of elytra, breast and apex of antennae more or less black. L. 1¾—2¼ l. Common. *H. minuta*, Lin.

B. Head, thorax and scutellum blackish, front and sides of thorax yellowish.

Ovate. Elytra brownish-yellow, suture, apex and outer margin blackish; antennae black, base yellow; under-side black, abdomen sometimes yellowish, tibiae and tarsi yellowish. L. 1¾—2 l. Moderately common.

H. marginata, Fab.

Microcara.

Oblong oval, slightly convex. Red yellow, scarcely shiny; not very closely covered with pale erect pubescence; antennae brownish yellow, paler at base. Thorax closely and very finely punctured; elytra with three scarcely raised lines, closely punctured. L. 2 l. Common.

M. livida, Fab.

M. Bohemani, Mann., is doubtfully distinct from *M. livida*; it differs in being smaller and more shiny, and in having the vertex of head, middle of thorax and middle joints of antennae black-brown, with the side margins of thorax more raised.

Cyphon.

A. Elytra with feeble longitudinal elevations.

a. Body moderately convex.

Nearly oval. Lighter or darker reddish-brown; base of antennae and legs yellow; pubescence rather strong, gray. Thorax narrowed in front; elytra evenly, closely and moderately strongly punctured. L. 1—1¼ l. Common.

C. coarctatus, Payk.

C. fuscicornis, Th. has the punctuation of elytra closer and finer round scutellum than elsewhere; first three joints of antennae and the legs red-yellow, rest of former brown.

b. Body not very convex.

I. Elytra evenly, not very closely and rather strongly punctured.

Oval. Blackish-brown ; base of antennae and legs yellow. More shiny than *C. coarctatus,* pubescence feebler, thorax more narrowed in front. L. 1 l. Rather common.

C. nitidulus, Th.

II. Elytra more closely punctured round scutellum than elsewhere.

Oval. Pale ochre-yellow ; apex of antennae, vertex of head and breast brown ; pubescence feeble. L. 1½ l. Moderately common.   *C. pallidiventris,* Th.

B. Elytra without trace of longitudinal elevations.

a. Length 1—1¼ lines.

I. Pubescence close; elytra finely punctured throughout.

Not very convex. Pale brownish-yellow, base of head, eyes, apex of antennae and hinder part of suture of elytra blackish ; under-side sometimes with dark spots ; pubescence gray. Elytra finely and closely punctured. Common.   *C. variabilis,* Thunb.

II. Pubescence scanty ; elytra strongly punctured at base.

Oval, convex, moderately shiny. Black; thorax, elytra and legs yellow-red ; base of antennae obscurely reddish ; pubescence very short, fine, scanty. Shorter, broader and more convex than *C. variabilis,* elytra more diffusely punctured, with punctuation at base coarser, third joint of antennae rather shorter. Elytra (but not suture only) sometimes brownish, lighter at apex than at base. L.1¼l. Rare.

C. punctipennis, Sharp (*nigriceps,* Sharp's Cat.)

b. Length ¾ line.

I. Fourth joint of antennae oblong,

Oval. Reddish-yellow ; eyes black ; antennae brownish at apex; pubescence rather feeble. More convex and shiny than *C. variabilis,* with elytra rather more strongly and less closely punctured. Common. *C. pallidulus,* Boh.

II. Fourth joint of antennae reversed conical.

Oval, slightly convex. Black ; base of antennae and apex of elytra, tibiae and tarsi red-brown ; margins of thorax and elytra sometimes lighter. Common.

C. padi, Lin.

Prionocyphon.

Roundish oval, rather convex. Reddish-yellow; pubescence yellow, moderately close and strong. Elytra strongly
and rather closely punctured. L. 1½—2 l. Rare.

P. serricornis, Müll.

Hydrocyphon.

Roundish-ovate. Brown; base of antennae and legs
paler. Head and thorax short, deflexed; elytra very convex in front, broadest before middle. Punctuation very
fine. L. ¾ l. Moderately common. *H. deflexicollis*, Müll.

Scirtes.

Broad oval. Black; base of antennae, tibiae and tarsi,
and sometimes sides of thorax yellowish. Thorax very
finely, elytra finely punctured; pubescence very feeble.
L. 1¼—1¾ l. Common. *S. hemisphaericus*, Lin.

EUBRIIDAE.

Antennae sawlike, second joint very small; labial palpi
three-jointed, last joint triangular with four blunt spines
at apex. Eyes large and oblong. Thorax double as broad
as long; elytra with longitudinal furrows. *Eubria*, Redt.

Eubria.

Roundish, convex, rather shiny. Blackish; base of antennae, tibiae and tarsi lighter. Punctuation extremely
fine, pubescence fine. Elytra with five furrows on each.
L. ¾—1 l. Scarce. *E. palustris*, Germ.

LYCIDAE.

Head small, not produced in front, covered by thorax;
inner lobe of maxillae obsolete, outer one broad and short,
scantily fringed; prosternum short; abdomen with seven
ventral segments. Thorax with raised sides and several
raised lines enclosing spaces. *Eros*, Newm.

E 2

Eros.

A. Each elytron with ten rows of punctures or longitudinal impressed lines.

a. Thorax (except in middle) scarlet; length 3½—4½ lines.

Black; margins of thorax and the elytra scarlet, latter with nine ribs (alternately stronger and weaker), with regular transverse wrinkles forming a network in the interstices. Not uncommon. *E. Aurora,* Fab.

b. Thorax black; length 2—3 lines.

Black; elytra scarlet, with four feeble ribs, each interstice with two rows of moderately large, closely placed punctures separated by a very feeble rib. Third joint of antennae double as long as second, their last two joints brownish-yellow. Rare. *E. minuta,* Fab.

B. Each elytron with four strong ribs, interstices shewing regular transverse ridges not intersected by any longitudinal raised line.

Black; elytra scarlet. Third joint of antennae not much longer than second. L. 3—3½ l. Rare. *E. affinis,* Payk.

LAMPYRIDAE.

A. Mandibles slender; prothoracic spiracles hidden.

Both sexes without wings; head covered by thorax; elytra of male very short, in female absent.

Phosphaenus, Cast.

B. Mandibles small, moderately broad; prothoracic spiracles distinct.

Male with wings and long elytra; head covered by thorax; female without either wings or elytra.

Lampyris, Geoffr.

Phosphaenus.

Blackish-brown; last two abdominal segments whitish-yellow (especially beneath); tibiae and tarsi brown; pubescence scanty. Antennae thick; elytra of male shorter than thorax. L. 3—3½ l. Rare. *P. hemipterus,* Geoffr.

Lampyris.

Male brownish-yellow; disc of thorax and elytra darker;

apex of abdomen pale. L. 5—6 l. Female brownish; margins of thorax and abdomen paler, last three segments yellow. L. 6—8 l. Common. *L. noctiluca*, Lin.

DRILIDAE.

Mandibles curved, with sharp apex and a sharp tooth in middle of inner-side; maxillae with only one, small lobe.

Drilus, Ol.

Drilus.

Male black; elytra yellow; pubescence rather coarse, yellow. Antennae comblike. L. 2½—3 l. Female without elytra or wings, antennae short. L. 6 l. Male moderately common, female rare. *D. flavescens*, Ol.

TELEPHORIDAE.

A. Last joint of palpi hatchet-shaped; elytra not or scarcely abbreviated.

   a. Mandibles not toothed on inner-side; sides of thorax not emarginate toward base.

Antennae thread-like. *Telephorus*, Schaeff.

   b. Mandibles toothed on inner-side; sides of thorax emarginate toward base in male and often also in female.

Antennae often feebly sawlike. *Silis*, Latr.

B. Last joint of palpi ovate; elytra abbreviated.

   a. Antennae inserted at some distance from eyes; mandibles with a tooth on inner-side.

Head prominent, only slightly inclined, strongly narrowed at base; elytra longer than in *Malthodes*.

Malthinus, Latr.

   b. Antennae inserted close to eyes; mandibles not toothed on inner-side.

Head bent downward, more or less narrowed at base.

Malthodes, Kies.

Telephorus.

A. Head strongly narrowed and constricted behind, forming a neck.

Apical spines of tibiae fine, often indistinct; claws split.
[*Podabrus*, Fisch.]

Head black, yellow before eyes; thorax yellow, with a black longitudinal spot, sometimes covering all the disc and anterior margin, at others indistinct; elytra yellow or black; under-side blackish, prosternum, margins and apex of abdomen yellow; antennae yellow, separate joints brownish at apex; legs yellow (sometimes black), tarsal joints black at apex. Head strongly and closely punctured, somewhat in wrinkles; thorax half as broad again as long, posterior angles toothed, often indistinctly; third joint of antennae scarcely longer than second and much shorter than fourth. L. 5—6 l. Moderately common.

T. alpinus, Payk.

B. Head not strongly narrowed and not (or only indistinctly) constricted behind.

    a. Apical spines of tibiae distinct; claws simple or toothed at base [*Telephorus*].

      I. Elytra dark blue; both claws of male with a prominence and of female with a fine spinelike tooth at base.

Black; male with front of head, prosternum and abdomen reddish-yellow, elytra dark blue; female with thorax and first two joints of antennae also red-yellow. L. 5—7 l. Scarce.   *T. abdominalis*, Fab.

      II. Elytra black or pale brownish-yellow; outer claw only more or less toothed.

        1. Length 5—6½ lines.

        A A. Thorax marked with black.

          a a. Thorax red-yellow, with a black spot (variable in extent) in middle of anterior margin; femora (except sometimes anterior pair) black.

Robust. Head, elytra and under-side black, with a gray reflection through pubescence; front of head, base of antennae, inner-side of anterior tibiae, margins and apex of abdomen red-yellow. Third joint of antennae half as long again as second and shorter than fourth; tooth on claws rather small and blunt. L. 5½—6½ l. Rather common.

T. fuscus, Lin.

b b. Thorax red-yellow, with a black spot in middle of disc; all femora red.

Similar to *T. fuscus* but rather narrower, antennae stronger, thorax rather flatter and tooth on claw rather larger. L. 5—6 l. Common. *T. rusticus*, Fall.

B B. Thorax unicolorous red-yellow.

Head reddish-yellow, with a black spot (not touching eyes) on forehead; elytra yellow; prosternum reddish-yellow, with a small brown plate between coxae, rest of under-side black, with more or less yellow on margin and apex of abdomen; antennae blackish, base lighter; legs reddish, on posterior pair upper-side of apex of femora, the tibiae and part of tarsi black, part of intermediate tibiae and tarsi also black. Fourth joint of antennae rather longer than third and from two to three times longer than second; front and sides of thorax evenly rounded; tooth on claws strong and sharp. L. 5—6 l. Common.

T. lividus, Lin.

The variety *T. dispar*, Fab. has the elytra black.

2. Length 3—5 lines.

A A. Elytra black; antennae with a smooth impressed line on central joints.

a a. Legs chiefly yellow.

A a. Third joint of antennae about twice as long as second joint.

Head black, reddish-yellow below eyes; thorax red-yellow; prosternum, abdomen and legs red-yellow, posterior tibiae and tarsi more or less pitch-brown, rest of breast black; antennae reddish-brown, lighter at base and darker at apex. Thorax flatly rounded in front and only slightly so at sides; tooth on claws spinelike. L. 4½—5 l. Common. *T. pellucidus*, Fab.

B b. Third joint of antennae about one-third longer than second joint.

Head black, from base of antennae forward yellow; thorax reddish-yellow, with a blackish spot (variable in extent) in middle; under-side reddish-yellow, meso- and metasternum, and sometimes some spots at sides of abdominal segments, black; antennae reddish-yellow, darker toward apex; legs red-yellow, on posterior pair apex of femora and tibiae until about middle usually blackish. Tooth on claws large. L. 5 l. Common.

T. *nigricans*, Müll.

T. discoideus, Steph., is a somewhat small variety of *T. nigricans*.

b b. Legs entirely black.

Black; mouth, sides of thorax and of abdomen and under-side of base of antennae yellow. Third joint of antennae about half as long again as second and much shorter than fourth ; tooth on claws large and sharp. L. 4—5 l. Moderately common. *T. obscurus*, Lin.

B B. Elytra yellow or pale yellow-brown ; or elytra black, and antennae without bare impressed lines.

a a. Antennae of male without bare impressed lines on central joints.

A a. Antennae of male long and slender, with third joint nearly twice as long as second.

Reddish-yellow ; eyes, metasternum and base of abdominal segments more or less blackish ; apex of antennae brownish ; vertex of head and disc of thorax sometimes dark ; legs varying from dark with light knees to unicolorous yellow. Male with antennae long, slender, simple, third joint nearly twice as long as second. Female with antennae rather long, third joint rather longer than second, seventh ventral segment of abdomen sinuate on each side, central lobe acutely incised at apex ; elytra almost shorter than abdomen. L. 4—4½ l. Common. *T. lituratus*, Fall.

B b. Antennae of male rather thick, with third joint not quite half as long again as second.

Black; base of antennae, front of head, thorax, and knees yellow-red ; thorax nearly square, diffusely and indistinctly punctured, with a variable black marking behind ; elytra and tibiae sometimes red-yellow, latter with black lines. Male with antennae rather thick, simple, third joint nearly half as long again as second, and elytra almost shorter than abdomen. Female with antennae shorter, third joint not much longer than second, and elytra shorter than abdomen ; seventh ventral segment of abdomen sinuate on each side, with central lobe acutely incised at apex. L. 4—5 l. Not common; Aberlady, under seaweed.

T. *Darwinianus*, Sharp.

b b. Antennae of male with smooth impressed lines on central joints.

A a. Head rather long and broad.

Reddish-yellow; vertex of head often blackish; elytra more or less brownish ; under-side (except margins of abdominal segments), knees of hinder pairs of legs and rarely of anterior pair blackish. Third joint of antennae double as long as second and scarcely shorter than fourth ; sides of thorax slightly, anterior margin more strongly rounded ; tooth on claws blunt. L. 3—3½ l. Common.

T. *bicolor*, Fab.

B b. Head small, much narrowed toward base.

A 1. Space on head between bases of antennae yellow.

Brown-yellow ; head (except front), a spot on thorax (sometimes absent), antennae (except base) and under-side blackish ; legs varying from smoky-brown, with apex of femora and tarsi lighter, to unicolorous yellow. Third joint of antennae fully half as long again as second and distinctly shorter than fourth ; eyes large and prominent ; thorax somewhat narrowed in front; outer claw on all tarsi with a distinct tooth. L. 3—3½ l. (? Brit.)

T. *figurata*, Mann.

B 1. Space on head between bases of antennae black.

Elongate. Black ; base of antennae, mouth, margins of thorax (more or less), elytra and tibiae red-yellow. Antennae of male with third joint nearly twice as long as second and an impressed line on joints four to ten ; those of female shorter, simple, third joint almost half as long again as second. Antennae longer and stouter than in T. *figurata*, thorax larger, with anterior angles and apical margin more rounded, the black colour spread over a greater area. Elytra and legs longer. L. 3—4 l. Rare.

T. *scoticus*, Sharp.

3. Length 2—3 lines.

A A. Elytra entirely yellow.

Upper-side yellow; basal half of head and a large spot on disc of thorax black; under-side black. Margins of abdominal segments and legs (except sometimes apex of posterior femora) yellow. Third joint of antennae half as long again as second and not much shorter than fourth; thorax not narrowed in front; claws thin, without any distinct tooth. L. 2½—3 l. Common.

T. haemorrhoidalis, Fab.

B B. Elytra entirely blackish, or black with yellow margin.

    a a. Posterior claws without tooth.

        A a. Thorax red-yellow, with margins lighter.

Head black, front yellow; elytra black, with yellow border, pubescence whitish, leaving some spots bare, and with a mixture of longer yellow hairs; under-side of head reddish-yellow, with black marking, breast and abdomen black, apex, sides and margins of segments of latter yellow; antennae yellow-red, brownish toward apex; legs reddish-yellow, posterior tibiae often more or less brown. Third joint of antennae not much longer than second and almost as long as fourth; margins of thorax nearly straight; all claws of female without teeth, of male anterior claws with a sharp tooth and intermediate pair with a weaker one. L. 2—2½ l. Common. *T. lateralis*, Lin.

        B b. Thorax blackish, with side margins yellowish in front.

Head blackish; elytra elongate, rather shiny, black, under-side and legs black, base and apex of tibiae yellowish, tarsi brown; antennae black, base reddish-yellow. Third joint of antennae half as long again as second and not much shorter than fourth; thorax much broader than long, sides rounded; male with anterior claws toothed and intermediate pair more feebly so, female with anterior claws toothed. L. 2½—3 l. Rather common. *T. paludosus*, Fall.

    b b. Posterior claws toothed.

        A a. Thorax unicolorous yellow.

Head pitch-brown, reddish-yellow in front; elytra pitch-brown, sometimes paler at suture; under-side and legs reddish-yellow, breast rarely more or less brown; antennae yellow, toward apex brownish. Third joint of antennae double as long as second and as long as fourth; sides of thorax tolerably straight, anterior margin rounded; outer claws on all tarsi with a moderately strong tooth. L. 2½ l. Moderately common. *T. thoracicus*, Gyll.

B b. Thorax pitch-brown, margins reddish.

Head black, yellow in front; elytra pitch-brown; under-side black, sides, apex and margins of segments of abdomen reddish-yellow; antennae pitch-brown, first three joints reddish-yellow; legs reddish-yellow, base of femora blackish. Third joint of antennae double as long as second and rather shorter than fourth; outer claw on all tarsi with a moderately large tooth. L. 2½—3 l. Moderately common. *T. flavilabris*, Fall.

b. Apical spines of tibiae indistinct; tarsal claws split.

Head small, moderately prominent; thorax somewhat narrowed in front; tibiae slender straight.
[*Rhagonycha*, Eschsch.]

I. Thorax entirely yellow or reddish-yellow.

1. Elytra unicolorous yellow.

Yellow; eyes very large and prominent, black; abdomen more or less spotted with black. Third joint of antennae twice as long as second joint; thorax rather broader than long, anterior margin somewhat rounded. L. 4½—5 l. Not common. *T. translucidus*, Kry.

2. Elytra yellow, with apex black.

A A. Head (except mouth) black.

Under-side black; prosternum, apex of abdomen and legs yellow; antennae black, base yellow. Third joint of antennae scarcely twice as long as second and not much shorter than fourth; thorax not much broader than long. L. 3—3½ l. Rather common. *T. fuscicornis*, Ol.

B B. Head yellow-red.

Eyes and under-side black; antennae brown or blackish, base lighter; legs red-yellow, tarsi darker. Third joint of antennae half as long again as second and much shorter than fourth; thorax longer than broad, much narrowed in front. L. 2½—4 l. Common. *T. fulvus*, Scop.

II. Thorax black, with sides or all margins yellow.

1. Sides of thorax, femora and anus yellow.

Short. Head black, mouth lighter ; elytra yellow; under-side black, anus and legs yellow ; antennae blackish, base yellow, second and third joints about equal in length. Thorax broader than long, apical margin somewhat rounded, anterior angles almost right angles. Male with penultimate ventral segment of abdomen widely emarginate at apex. L. 2 l. Common.   *T. testaceus*, Lin.

2. All margins of thorax yellow ; femora and anus black.

Black, base of antennae, side margins of thorax broadly, its apical and basal margins narrowly, elytra and legs yellow, latter with femora black. Similar to *T. testaceus* but with apical margin of thorax truncate, anterior angles right angles and penultimate ventral segment of abdomen of male depressed at apex, nearly truncate and produced into a tooth on each side. L. 2 l. Common. *T. limbatus*, Th.

III. Thorax unicolorous black.

1. Elytra yellow.

Elongate. Head (except mouth) black ; under-side and antennae black, base of latter and the legs yellow. Thorax small, about as long as broad, distinctly narrowed in front. L. 2½—3 l. Common.   *T. pallidus*, Fab.

2. Elytra black.

Elongate. Black ; first two joints of antennae and knees yellowish. Thorax about as long as broad, somewhat narrowed in front. L. 2½ l. Moderately common.

T. elongatus, Fall.

Silis.

Head black, mandibles red-yellow ; thorax yellow-red ; elytra black ; prosternum and abdomen reddish-yellow, rest of breast and legs black, latter sometimes lighter. Thorax transverse, uneven, irregularly punctured ; antennae strong, slightly sawlike, black. L. 2½—3 l. Moderately common.   *S. ruficollis*, Fab.

Malthinus.

A. Length 1½—1¾ lines.

a. Elytra tolerably regularly punctured in rows, more

than twice as long as their united breadth, apex sulphur-yellow.

I. Posterior legs yellow.

Head black behind, reddish-yellow in front ; thorax black, sides reddish-yellow ; elytra gray-yellow, a badly defined spot round scutellum and a broad transverse band before the sulphur-yellow apex blackish ; under-side, base of antennae and legs yellow, breast more or less dark. Thorax about as long as broad. Rather common.

M. fasciatus, Fall.

II. Posterior legs brown.

Brown ; forehead, apex of elytra, base of antennae and front legs yellow ; elytra with a broad pale transverse band on front half. Very similar to *M. fasciatus,* but smaller, darker and with longer thorax. Rather common.

M. balteatus, Suf.

b. Elytra confusedly punctured, somewhat in wrinkles, more than three times as long as their united breadth, unicolorous black.

Black ; mouth, base of antennae and legs reddish ; front of head lighter in male ; thorax rather broader than long. Not uncommon. *M. frontalis,* Marsh.

B. Length $2\frac{1}{4}$—$2\frac{3}{4}$ lines.

Head black behind, yellow in front; thorax yellow, with two black spots, only indicated in male, often extensive in female ; scutellum yellow ; elytra gray-yellow, base, suture and a spot before the sulphur-yellow apex dark ; abdomen and legs yellow, under-side of head and breast blackish. Head with a furrow at base ; elytra not much shorter than abdomen, rather finely punctured, with indistinct traces of striae. Common. *M. punctatus,* Fourc.

Malthodes.

A. Length 1 line or more.

a. Head (without eyes) as broad as, or broader than thorax.

I. Antennae entirely blackish or brownish.

1. Thorax blackish, with base, apex and more or less of hinder part of side margin yellow ; length $2\frac{1}{4}$—$2\frac{1}{2}$ lines.

Grayish-brown; mouth, apex of elytra (not very sharply defined) and some markings on under-side yellow; legs yellowish-brown, femora rather darker. Head strongly narrowed behind, finely and closely punctured, impressed between eyes; thorax about as broad as (male), or rather broader than long (female), posterior angles not very sharp right angles; elytra not quite three times as long as together broad. Male with last dorsal abdominal segments scarcely elongate, simple, penultimate ventral segment roundly emarginate, last ventral segment very narrow, rather broader at base, split almost to base, the two portions slightly curved from one another. Common.

M. marginatus, Latr.

2. Thorax black, with a narrow yellow border at base ; length 1½—1¾ lines.

Black; apex of elytra (more or less distinctly), sides and apex of abdomen yellow; legs brownish-gray, knees rather lighter. Head narrowed behind, with an inconspicuous impression between eyes; thorax much broader than long, posterior angles rather sharp right angles ; elytra scarcely 2½ times as long as together broad. Male with last dorsal abdominal segments moderately elongate, the last one somewhat dilated toward apex, front margin straight, with a triangular excision in middle ; penultimate ventral segment deeply emarginate; last ventral segment deeply split, laterally compressed, with angular emarginations. Rather common.   *M. mysticus*, Kies.

II. Antennae pitch-brown, first two joints reddish-yellow.

Gray-black or brownish ; margins of thorax, base of tibiae, tarsi and markings on under-side yellowish. Head somewhat narrowed behind ; thorax not much broader than long, posterior angles rounded ; elytra about 2½ times as long as together broad. Male with penultimate dorsal abdominal segment deeply emarginate, lobes with an elbowed process, last dorsal segment small, penultimate ventral segment moderately deeply emarginate, last ventral segment somewhat narrowed in middle, rather deeply triangularly excised at apex. L. 1½ l. Not common.

M. fibulatus, Kies.

b. Head (without eyes) narrower than thorax.

I. Thorax broader than long, posterior angles not rounded.

1. Elytra almost three times as long as together broad.

Gray-brownish; apex of elytra and part of under-side yellow; antennae brownish, base paler; legs brownish-yellow, femora darker. Head not much narrowed behind, diffusely punctured, without furrow between eyes; elytra almost double as broad as thorax. Male with last dorsal abdominal segments not elongate, simple; penultimate ventral segment deeply triangularly excised, last ventral segment narrow, scarcely divided at apex. L. 1½—2 l. Rather common. *M. pellucidus*, Kies.

2. Elytra about 2½ times as long as together broad.

A A. Elytra pitchy or brownish-black, with apex yellow.

a a. Thorax pitch-black or pitch-brown.

A a. Antennae reaching somewhat beyond apex of elytra.

Pitch-black or pitch-brown; mouth, base of antennae, apex of elytra, part of under-side and legs yellow, femora often rather darker. Head not much narrowed behind, rather finely and feebly punctured, with a shallow impression between eyes; anterior margin of thorax tolerably straight in middle, oblique on each side. Male with antepenultimate dorsal abdominal segment produced into a curved hook at anterior angles, penultimate dorsal segment with a simple, rather sharp process, penultimate ventral segment deeply roundly emarginate, last ventral segment somewhat divided at apex. L. 2—2¼ l. Not uncommon.
M. dispar, Germ.

B b. Antennae as long as, or longer than whole of body.

Pitch-black; mouth, apex of elytra and part of under-side yellow; base of antennae sometimes light; legs brownish, tibiae rather darker. Head distinctly narrowed behind, rather finely and closely punctured; anterior margin of thorax gently rounded. Male with antepenultimate dorsal abdominal segment produced into a rather sharp point at anterior angles, penultimate dorsal segment with a

shorter and blunter point, penultimate ventral segment
slightly emarginate, last ventral segment rather narrow,
flattened, tolerably parallel-sided, grooved throughout its
length and triangularly emarginate at apex. L. $1\frac{3}{4}$—2 l.
Moderately common.   *M. flavoguttatus*, Kies.

    b b. Thorax yellow-red, often blackish in
       middle.

Head brown, mouth reddish-yellow ; under-side brownish-
black, with yellow markings, antennae pitch-brown, base
reddish-yellow ; legs reddish-yellow, femora blackish. Head
scarcely narrowed behind, very finely punctured, with a
more or less distinct furrow between eyes ; anterior margin
of thorax rounded. Male with last dorsal abdominal segments simple and not elongate, penultimate ventral segment
moderately deeply emarginate, last ventral segment somewhat elongate, split to base. L. $1\frac{1}{4}$—$1\frac{3}{4}$ l. Common.
                          M. sanguinolentus, Fall.

    B B. Elytra unicolorous black.

Black ; base of antennae rarely lighter ; legs dark brown,
knees rather lighter. Head narrowed behind, extremely
finely and diffusely punctured, impressed between eyes ;
thorax double as broad as long, all angles raised and slightly
prominent ; antennae not reaching apex of elytra. Male
with penultimate dorsal abdominal segment somewhat
elongate, simple, last dorsal segment very narrow, apparently ·
composed of two pieces placed side by side, penultimate
ventral segment deeply emarginate, last ventral segment
somewhat short, divided at apex. · L. 1—$1\frac{1}{4}$ l. Rare.
                          M. nigellus, Kies.

    3. Elytra about twice as long as together broad.

Black ; thorax generally with narrow yellow border at
base and apex ; elytra with yellow apex ; under-side with
yellow markings ; antennae pitch-brown, base often lighter ;
legs brown. Head somewhat narrowed behind, very finely
punctured ; elytra with traces of fine raised lines. Male
with last dorsal abdominal segments elongate, penultimate
one with posterior margin produced in middle, projecting
over last segment, which is bent downward, somewhat impressed in middle and emarginate at apex, penultimate
ventral segment deeply triangularly emarginate, last ventral
segment split to base. L. $1\frac{3}{4}$—2 l. Rare.
                          M. guttifer, Kies.

II. Thorax not broader than long, posterior angles rounded.

Black; part of under-side yellow; legs brown, knees lighter. Head not much narrowed behind, finely punctured, without frontal furrow; elytra more than $2\frac{1}{2}$ times as long as together broad. Male with penultimate dorsal abdominal segment elongate, emarginate at apex, produced on each side into a not very long lobe, suddenly elbowed at apex, penultimate ventral segment emarginate, last ventral segment very narrow, wider at base and apex, which is almost truncate. L. $1\frac{1}{2}$—$1\frac{3}{4}$ l. Rare. *M. misellus*, Kies.

B. Length $\frac{1}{2}$—$\frac{3}{4}$ lines.

Yellowish-gray; apex of elytra and sometimes margins of thorax dirty yellow; antennae brownish, scarcely reaching apex of elytra. Thorax more than double as broad as long, all angles blunted; elytra $2\frac{1}{2}$ times as long as together broad. Male with last dorsal abdominal segments elongate, apical ones deeply split, penultimate ventral segment roundly emarginate and produced on each side into long, pointed lobes, last ventral segment strongly curved, somewhat dilated at base, forked at apex. Rather common.

M. atomus, Th.

MELYRIDAE.

A. Episterna of mesothorax not separate from sternum; metasternum prominent behind.

Insects with retractile vescicles at anterior angles of thorax and beside first abdominal segment. *Malachiides.*

B. Episterna of mesothorax separate from sternum; metasternum truncate behind.

Insects without retractile vescicles. *Dasytides.*

MALACHIIDES.

A. Antennae inserted between eyes.

Ligula straight in front; labrum nearly as long as broad, slightly rounded in front. *Malachius*, Fab.

B. Antennae inserted on sides of head.

Ligula rounded in front; labrum much broader than long, truncate in front. *Anthocomus*, Er.

Malachius.

A. Thorax green, at most with anterior angles red.

a. Elytra red, with a common green longitudinal band, narrowed and abbreviated behind.

Head and under-side green, front of former yellow; anterior angles of thorax red.   L. 3—3½ l.   Common.
M. aeneus, Lin.

b. Elytra green, with a red spot at apex.

I. Thorax much broader than long, anterior angles bordered with red ; length 2½—3 lines.

Head and under-side green, front of former yellow. Third joint of antennae short; in male the second, third and fourth joints dilated on inner-side.   Common.
M. bipustulatus, Lin.

II. Thorax as broad as long, unicolorous green ; length 2 lines.

Head and under-side green, front of former yellow ; elytra sometimes unicolorous green.   Third joint of antennae moderately elongate, second, third and fourth joints in male simple.   Male with apex of anterior tibiae yellowish.   Rather common.   *M. viridis*, Fab.

B. Thorax green or blackish, with a red side border.

a. Body green ; length 2½ lines.

Green ; mouth yellow, sides of thorax and apex of elytra red.   Anterior tarsi simple.   Male with third to seventh joints of antennae emarginate beneath and produced on inner-side.   Moderately common.   *M. marginellus*, Ol.

b. Body black-green ; length 1½—1¾ lines.

Black-green ; mouth yellow; sides of thorax and apex of elytra red-yellow.   Male with second joint of anterior tarsi obliquely produced at apex and studded with short bristles. Moderately common.   *M. pulicarius*, Fab.

C. Thorax entirely red.

Black-green ; thorax and apex of elytra red.   Male with second joint of anterior tarsi obliquely produced at apex and studded with short bristles.   L. 1½ l.   Moderately common.
M. ruficollis, Ol.

Anthocomus.

A. Elytra entirely red.

Black-green; elytra and sides of thorax red. L. 2 l. Not uncommon. *A. sanguinolentus*, Fab.

B. Elytra partly black or green.

a. Thorax green.

Head green; elytra black, with a transverse band (interrupted at suture) before middle and the apex red; under-side black. L. 1⅔ l. Common. *A. fasciatus*, Lin.

b. Thorax yellow-reddish.

Head black; elytra green, apex (sometimes nearly to middle) yellow; under-side green; tibiae and tarsi reddish-yellow. L. 1½ l. Not uncommon.

A. terminatus, Men.

DASYTIDES.

A. Antennae saw-like.

a. Tarsal claws without membranous appendages.

Maxillary palpi threadlike; tarsi generally elongate, first joint (especially of posterior pair) usually longer than last one. *Dasytes*, Payk.

b. Tarsal claws dissimilar, one having a membranous appendage attached to it throughout, the other only as far as middle.

Maxillary palpi threadlike; tarsi more or less elongate, first joint as long as, or longer than last one.

Dolichosoma, Steph.

c. Tarsal claws with a free membranous appendage, attached at base only, but reaching as far as apex.

Maxillary palpi somewhat thickened, last joint truncate at apex; tarsi generally short and thick, first joint shorter than last one. *Haplocnemus*, Steph.

B. Antennae bead-like, with three larger apical joints.

Inner maxillary lobe, with a horny hook at apex; maxillary palpi short, last joint pointed at apex; tarsi long and slender, first joint shorter than second; claws simple.

Phloeophilus, Steph.

Dasytes.

A. Thorax as long as broad; elytra with scabrous punctuation.

a. Thorax without central furrow, sides parallel; fourth joint of tarsi much shorter and narrower than third.

I. Tibiae red-yellow.

Elongate linear. Leaden-greenish, with yellow-gray pubescence and brown hairs ; second joint of antennae, almost whole of anterior legs and trochanters, tibiae and tarsi of hinder pairs red-yellow. Thorax closely and finely punctured. L. 2 l. Rare. *D. oculatus*, Kies.

II. Tibiae leaden-black.

Elongate linear. Leaden-green, with pale gray pubescence and brown hairs; trochanters and base of posterior tarsi red-yellow. Thorax with disc diffusely punctured, sides wrinkled. L. 2 l. Common.
D. plumbeo-niger, Goez.

b. Thorax with fine central furrow, sides somewhat narrowed in front; fourth joint of tarsi only slightly shorter and narrower than third.

Elongate. Bronze-black, with pale gray pubescence and brown hairs; trochanters, tibiae and tarsi red-yellow, latter brown at apex. Thorax shorter than in *D. oculatus* and *plumbeo-niger*, almost transverse, with posterior angles somewhat obtuse. L. 2 l. Common. *D. plumbeus*, Müll.

B. Thorax slightly transverse ; elytra very finely and closely punctured.

Rather broad. Black, hairy, moderately shiny. Thorax tolerably strongly, not very closely punctured. L. 2 l. Rare. *D. niger*, Lin.

Dolichosoma.

A. Antennae rather short, body covered with hairs.

a. Head and thorax closely punctured in wrinkles; body rather dull.

Green or blue; with fine gray pubescence mixed with stronger black hairs. Thorax as long as broad, with a more or less distinct central furrow ; elytra with traces of raised longitudinal lines. L. 2½—3 l. Common. *D. nobilis*, Ill.

b. Head and thorax closely covered with punctures flat at the bottom, which is again punctured in middle; body shiny.

Green or blue; with scattered, long, erect hairs. Thorax elongate; elytra somewhat dilated behind. L. 2—2½ l. Isle of Wight. *D. protensa*, Gen.

B. Antennae tolerably long; body covered with oblong scalelike hairs.

Elongate. Greenish-bronze, dull, with a grayish appearance on account of the scalelike hairs. Thorax double as long as broad; upper-side very closely and rather finely punctured. L. 2½ l. Moderately common.

D. linearis, Rossi.

Haplocnemus.

A. Body greenish-blue.

Oblong, convex, scantily covered with pale hairs, greenish-blue, base of antennae, tibiae, tarsi and sometimes femora and reflexed margin of elytra red-yellow. Thorax finely and not very closely punctured, narrower than elytra, which are diffusely and deeply punctured, the punctures being fewer than in *H. impressus*, coarser and less clearly defined. L. 2 l. Rare. *H. nigricornis*, Fab.

B. Body bronze-black.

Oblong. Bronze-black, covered with yellowish-hairs; tibiae either black, with base and apex pale, or entirely pale; tarsi pale beneath. Punctuation very close and coarse, somewhat confluent and irregular on elytra. L. 2 l. Not common. *H. impressus*, Marsh.

Phloeophilus.

Oblong. Dark brown or blackish, base of antennae and legs rather lighter; elytra yellowish, with interrupted waved brown spots and dark side margins; pubescence gray; punctuation rather strong. L. 1½ l. Not uncommon. *P. Edwardsi*, Steph.

CLERIDAE.

A. Tarsi distinctly five-jointed.

Antennae sawlike from third joint onward; labrum

rounded in front; last joint of maxillary palpi long oval, narrowed at apex; last joint of labial palpi strongly hatchet-shaped; tarsal claws with a tooth at base and apex. *Tillus*, Ol.

B. Tarsi indistinctly five-jointed, first joint being small and hidden.

a. Eyes strongly granulate and prominent.

Antennae with three larger apical joints; labrum emarginate in front; last joint of both pairs of palpi long hatchet-shaped; first joint of tarsi very short, next three with emarginate soles, claws simple. *Opilus*, Latr.

b. Eyes finely granulate, not strongly prominent.

I. Maxillary palpi threadlike.

Antennae gradually thickened; labrum emarginate in front; last joint of labial palpi large, hatchet-shaped; first joint of tarsi very short, next three with emarginate soles, claws simple or toothed at base. *Clerus*, Geoffr.

II. Maxillary palpi long hatchet-shaped.

Antennae with joints two to eight gradually shorter and broader, last three forming a moderately large and loosely articulated club; last joint of labial palpi a little broader than that of maxillary palpi; first joint of tarsi tolerably distinct, next three with very short, entire soles.
Tarsostenus, Spin.

C. Tarsi four-jointed (fourth joint being absent); first joint short.

Antennae with three larger apical joints; labrum deeply emarginate in front; abdomen with five ventral segments; tarsal claws toothed at base. *Corynetes*, Herbst.

Tillus.

A. Elytra entirely bluish-black; length 3½—4 lines.

Head and under-side black; thorax of male black, of female red; pubescence fine. Elytra with punctured striae. Not uncommon. *T. elongatus*, Lin.

B. Elytra black, with base red and a whitish-yellow, slightly curved band behind middle; length 2½—3 lines.

Head, thorax and under-side black; pubescence moderately close and long. Punctured striae on elytra reaching from base beyond middle, apex scantily punctured. Scarce. *T. unifasciatus*, Fab.

Opilus.

Rather flat. Dark brown; mouth, antennae, legs (except apex of femora) pale brownish-yellow, elytra with a spot (often divided) at shoulders, a transverse band close behind middle and the apex of this colour ; pubescence somewhat woolly, mixed with longer hairs. Elytra moderately closely and finely punctured, with some distinct but irregular rows (disappearing before apex) of larger punctures and traces of raised lines. L. 4—5 l. Not uncommon. *O. mollis*, Lin.

Clerus.

Head black ; thorax red, with anterior margin black ; elytra black, red at base, with a narrow, toothed line of white pubescence before middle and another broader line before apex ; under-side red, legs black. L. 3—4 l. Common. *C. formicarius*, Lin.

Tarsostenus.

Brown-black, hairy ; elytra with a transverse white band, somewhat behind middle ; antennae dark, with base dull red ; legs dull red, with base of femora black. Punctuation deep ; thorax with a central furrow. L. 2½—3 l. Rare.
T. univittatus, Rossi.

Corynetes.

A. Apical joint of palpi almost hatchet-shaped ; club of antennae with joints distinctly separate.

Blue ; antennae and legs black. Antennae as long as head and thorax, ninth joint more than double as broad as eighth, as long as broad, tenth not much larger than ninth ; elytra about twice as long as broad, with tolerably regular rows of fine, oblong punctures. L. 2 l. Common.
C. coeruleus, De G.

B. Apical joint of palpi cylindrical ; club of antennae with joints closely articulated.

a. Thorax red.

Head, antennae, abdomen and legs black ; elytra dark

blue, base red ; breast red. Thorax broader than long ;
elytra with distinct and regular, but rather fine punctured
striae, effaced before apex.   L. 2½ l.   Common.

C. *ruficollis*, Fab.

b. Thorax blue.

I. Antennae and legs red.

Dark blue.   Thorax much broader than long ; elytra
with irregular punctured striae, effaced somewhat behind
middle.   L. 2—2½ l.   Common.   C. *rufipes*, Fab.

II. Antennae and legs black.

Blue.   Antennae shorter than head and thorax, ninth
joint fully double as broad as, but not longer than eighth,
tenth much broader than ninth ; elytra less than twice as
long as broad, coarsely punctured in striae, interstices
punctured. L. 2 l. Moderately common. C. *violaceus*, Lin.

LYMEXYLONIDAE.

A. Antennae short, sawlike or (in male) comblike.

Maxillary palpi gradually a little thickened toward apex
in female, with appendages in male ; anterior coxae standing apart ; abdomen with seven ventral segments.

Hylecoetus, Latr.

B. Antennae rather long, threadlike.

Mouth parts as in *Hylecoetus ;* anterior coxae approximated ; abdomen with six ventral segments.

Lymexylon, Fab.

Hylecoetus.

Cylindrical ; antennae sawlike.   Male black ; legs and
sometimes elytra (except apex) yellowish.   Second joint
of maxillary palpi with two appendages, one of them
simple, the other double and bearing many fringed
branches.   L. 3—5 l.   Female rust-yellow, eyes and part
of breast black.   L. 1 –7 l.   Not uncommon.

H. dermestoides, Fab.

Lymexylon.

Elongate, cylindrical, pubescent.   Male black ; base of
elytra to middle of suture, abdomen, legs and base of antennae yellow.   Female ochre-yellow ; head, outer margin
and apex of elytra black. L. 2½—4 l.   Rare. *L. navale*, Lin.

TEREDILIA.

A. First tarsal joint about equal in length to second.

Abdomen with five ventral segments.

   a. First abdominal segment not much longer than the rest.

Antennae with from eight to eleven joints; prosternum without any process toward mesosternum; tarsi five-jointed.   *Ptinidae.*

   b. First abdominal segment much longer than the rest.

Antennae ten-jointed; prosternum without process toward mesosternum; tarsi five-jointed.   *Coniporidae.*

B. First tarsal joint very small.

Abdomen with five ventral segments, first segment much longer than the rest.

   a. Tarsi five-jointed.

     I. Club of antennae three-jointed.

Antennae with from nine to eleven joints, the three apical joints being larger; prosternum without process toward mesosternum; tarsi with five (rarely four) joints.   *Bostrychidae.*

     II. Club of antennae two-jointed.

Antennae eleven-jointed; elytra with punctured striae; second tarsal joint not large.   *Lyctidae.*

   b. Tarsi four-jointed.

Antennae with from eight to eleven joints, club three-jointed; tibiae without apical spines.   *Cissidae.*

PTINIDAE.

A. Antennae inserted on forehead.

Pronotum not separate from sides of prothorax.
                       Ptinides.

B. Antennae inserted close to anterior margin of eyes.

Pronotum separated by a raised line from sides of prothorax.   *Anobiides.*

PTINIDES.

A. Elytra punctured and hairy.

a. Third and fourth tarsal joints double as broad as long, emarginate at apex ; intermediate coxae contiguous.

Antennae eleven-jointed ; tarsal claws small.

Hedobia, Sturm.

b. Third and fourth tarsal joints simple ; intermediate coxae placed apart.

I. Scutellum distinct.

Bases of antennae almost contiguous ; elytra with regular and distinct punctured striae ; fifth tarsal joint only a little longer than second.   *Ptinus,* Lin.

II. Scutellum not distinct.

Bases of antennae placed apart ; eyes small ; elytra with indistinct striae ; fifth tarsal joint much longer than second.

Niptus, Boield.

B. Elytra smooth and shiny.

a. Thorax pubescent, with longitudinal ridges.

Head hairy ; antennae shorter than in *Gibbium,* last joint ovoid ; pointed at apex ; posterior trochanters moderately long.   *Mezium,* Leach.

b. Thorax bare, even.

Head not hairy ; last joint of antennae long, pointed at apex ; posterior trochanters nearly as long as femora.

Gibbium, Scop.

Hedobia.

Gray-brown ; pubescence on sides of thorax gray and on scutellum white ; elytra brown, with a spot at shoulder, the apex and a broad transverse band (dilated in middle and at each side) behind middle thickly covered with gray-white hairs. Thorax with a prominence before scutellum. L. 2 l. Not uncommon.   *H. imperialis,* Lin.

Ptinus.

A. Elytra of both sexes oblong, with parallel-sides and tolerably right-angled shoulders.

a. Elytra without markings.

Red-brown, sprinkled all over with white and yellowish

hairs; antennae and legs lighter. Thorax coarsely punctured, with four tubercles, the two inner ones pointed; elytra punctured in rows. L. 1½—2 l. Scarce.

P. germanus, Fab.

b. Elytra marked with white.

Brown; forehead, scutellum, a large spot behind shoulder of each elytron and another (usually double) before apex white; under-side with close gray-white pubescence. L. 1½ l. Not uncommon.

P. sexpunctatus, Panz.

B. Elytra (at least of female) rounded at sides.

a. Elytra with white spots.

I. Thorax with very fine central furrow.

Black or dark brown; scutellum white; elytra indistinctly clouded and waved with white; legs and antennae red. Thorax very convex, with two indistinct teeth; elytra with striae. Abdomen of female inflated. L. 1¼— 1¾ l. Rather common.

P. lichenum, Marsh.

II. Thorax with strong central furrow at base.

1. Elytra with a short, oblique line of points of white hair at shoulder; sides rounded in both sexes.

Brown or pitchy; antennae and legs lighter. Thorax globular, constricted behind, with four tubercles, the two inner ones blunt, central furrow not shiny at bottom and not fringed with white hairs at base; elytra with strongly punctured furrows, bristles longer than in *P. fur.* L. 1— 1¼ l. Rare.

P. subpilosus, Müll.

2. Elytra with two transverse bands of white hairs (at least in female); in male almost cylindrical, in female rounded at sides.

Red, red-brown or pitchy. Head with white hairs and a feeble central furrow; thorax strongly constricted before base, sides with a little tooth behind middle, disc with four tubercles, the two inner ones covered with oblong tufts of yellow hairs, convergent behind, central furrow shiny at bottom, fringed with white hairs at base; elytra with punctured striae. L. 1¼—1½ l. Common.

P. fur, Lin.

b. Elytra unicolorous; in male oblong, with parallel sides, in female oblong oval.

I. Thorax broader than long.

Rust-red; antennae and legs lighter. Thorax almost

globular, broader than long, strongly constricted behind (especially in female), closely covered with hair, with four distinct, conical tubercles near each other; elytra with strong punctured furrows and closely covered with hair. L. 1¼—1½ l. Rare.   *P. testaceus*, Ol.

II. Thorax oblong.

Yellow-brown or red-brown; antennae and legs lighter. Thorax oblong, constricted behind, bristly, with four equal, blunt tubercles; elytra with notched furrows. L. 1½—1¼ l. Rare.   *P. latro*, Fab.

Niptus.

A. Elytra with indistinct striae; length 1¾—2 lines.

Brown, closely covered with yellow pubescence. Thorax without tubercles; elytra globose ovate, interstices with rows of yellow erect hairs. Common. *N. hololeucus*, Fald.

B. Elytra with notched striae; length 1 line.

Dark red-yellow, rather dull, closely covered with pale gray pubescence and hairs. Thorax without tubercles; elytra globose ovate, interstices closely pubescent, with rows of short bristles. Rather common. *N. crenatus*, Fab.

Mezium.

Head and thorax clothed with whitish scales; elytra chestnut-brown, very shiny; antennae and legs whitish. Thorax with two central ridges and posterior margin thickened. L. 1¼—1¾ l. Rather common. *M. affine*, Boield.

Gibbium.

Chestnut-brown, shiny; antennae and legs closely covered with yellow, shiny hairs. Thorax very short. L. 1¼—1½ l. Not uncommon.   *G. scotias*, Fab.

ANOBIIDES.

A. No cavities on metasternum and abdomen for reception of hinder pairs of legs.

    a. Antennae eleven-jointed, threadlike, last three joints very long.

      I. Thorax obtuse at sides.

        1. Antennae placed near each other.

Thorax not convex ; elytra somewhat convex, strongly rounded at apex ; anterior coxae approximated, intermediate pair placed slightly, and posterior pair far apart ; ventral abdominal segments free, first slightly bisinuate at apical margin. *Dryophilus*, Chevr.

2. Antennae placed apart.

Thorax not convex ; elytra rather flat, obtuse at apex ; front pairs of coxae placed more or less apart, posterior pair far apart ; ventral abdominal segments free, first slightly bisinuate at apical margin. *Priobium*, Mots.

II. Thorax with a more or less prominent sharp edge.

1. Breast with a depression for reception of head ; elytra with striae.

Thorax more or less convex and uneven ; front pairs of coxae placed more or less, posterior pair moderately far apart ; ventral abdominal segments rarely soldered in middle, first more or less bisinuate at apical margin.
Anobium, Fab.

2. Breast without depression for reception of head ; elytra confusedly punctured.

A A. Anterior coxae placed somewhat apart.

Thorax not convex ; ventral abdominal segments free, first scarcely bisinuate at apical margin. *Xestobium*, Mots.

B B. Anterior coxae approximated.

Thorax not convex ; ventral abdominal segments free, first slightly sinuate at apical margin. *Ernobius*, Th.

b. Antennae eleven-jointed, sawlike or comblike, last three joints usually not, or scarcely larger than the rest.

I. Thorax without depression beneath for reception of head.

1. Antennae comblike or fanlike.

Lateral ridge of thorax fine ; intermediate coxae not farther apart than anterior pair ; posterior coxae not dilated inward ; first ventral abdominal segment slightly bisinuate at apical margin ; body elongate. *Ptilinus*, Geoffr.

2. Antennae sawlike.

Lateral ridge of thorax prominent ; intermediate coxae placed farther apart than anterior pair ; posterior coxae

dilated inward; first ventral abdominal segment almost
straight at apex ; body oblong. *Ochina*, Steph.

II. Thorax with depression beneath for reception of
head.

1. Metasternum simple.

Last joint of palpi more or less dilated ; intermediate
coxae placed apart; outer apical angle of posterior coxae
rounded ; first ventral abdominal segment nearly straight
at apex ; elytra with striae. *Xyletinus*, Latr.

2. Metasternum with a transverse ridge in front.

Last joint of palpi elongate; intermediate coxae approximated ; outer apical angle of posterior coxae sharp ; first
ventral abdominal segment nearly straight at apex ; elytra
confusedly punctured. *Lasioderma*, Steph.

B. Metasternum and base of abdomen with cavities for
reception of hinder pairs of legs.

α. Last three joints of antennae larger than the rest,
triangular, flattened.

I. Body oval.

Antennae ten-jointed ; eyes distinctly emarginate ; apex
of prosternum forked; metathoracic epistorna nearly parallel-sided ; posterior coxae dilated outward.

Dorcatoma, Herbst.

II. Body nearly hemispherical.

1. Antennae nine-jointed.

Eyes deeply emarginate ; apex of prosternum simple ;
metathoracic epistorna narrow at base, dilated behind ;
posterior coxae much dilated outward. *Coenocara*, Thoms.

2. Antennae eight-jointed.

Eyes scarcely at all emarginate ; apex of prosternum
forked ; metathoracic epistorna narrowed in middle ; posterior coxae nearly parallel-sided. *Anitys*, Thoms.

b. Last three joints of antennae larger than the rest,
rounded.

Antennae ten-jointed ; mandibles strong ; clypeus distinct from forehead ; body oblong; elytra as broad as
thorax. *Sphindus*, Chevr.

Dryophilus.

A. Scutellum not pubescent ; first ventral abdominal segment slightly produced in middle of apical margin.

Oblong cylindrical. Black or pitch-black; antennae and legs brown. Thorax short; elytra with almost indistinctly punctured striae, interstices very finely and closely punctured; antennae with joints three to eight distinctly longer than broad. L. ¾—1 l. Moderately common.

D. pusillus, Gyll.

B. Scutellum pubescent; first ventral abdominal segment strongly produced in middle of apical margin.

Elongate. Black, dull, anterior margin of thorax, apex and shoulders of elytra, antennae and legs red-brown. Thorax rather longer than broad; elytra with finely but distinctly punctured striae, interstices very finely and closely punctured; antennae with joints three to eight as broad as, or broader than long. L. 1—1½ l. Not common.

D. anobioides, Chevr.

Priobium.

Oblong cylindrical. Dark, dull chestnut-colour, with short yellow, shiny pubescence. Thorax broader than long; scutellum pubescent; elytra with strongly punctured striae, interstices convex, closely and finely punctured; antennae with joints three to eight scarcely longer than broad. L. 2½—3¾ l. Rare. *P. castaneum*, Fab.

Anobium.

A. Third joint of antennae about equal to second.

a. Thorax with right angled posterior angles and without tubercle.

Dark brown, dull; thorax with two patches of yellow hairs at base. Elytra with rather feeble punctured striae; interstices slightly convex, alternate ones raised at base; ventral abdominal segments soldered in middle; tarsi short. L. 2¼ l. Scarce. *A. denticolle*, Panz.

b. Thorax with obtuse posterior angles and with a tubercle toward base.

I. Mesosternal impression prolonged to middle of metasternum.

Lighter or darker pitch-brown; antennae and legs more or less reddish. Thoracic tubercle not enclosed by a horseshoe-shaped impression behind. Elytra with shallow but rather strongly punctured striae; interstices flat, alternate ones sometimes slightly convex at base; ventral abdominal segments free; tarsi somewhat long, very slightly thickened toward apex. L. 1½—2 l. Common.

A. domesticum, Fourc.

II. Mesosternal impression scarcely prolonged on to metasternum.

Black, dull; antennae, tibiae and tarsi reddish. Antennae rather shorter than in *A. domesticum*, elytra less rounded at apex, alternate interstices never slightly convex at base, tarsi more distinctly thickened toward apex. L. 1¾—2 l. Rather common.

A. fulvicorne, Sturm.

B. Third joint of antennae much shorter than second.

Short, oblong oval. Reddish-brown; antennae and legs lighter, eyes black. Thorax much broader than long, without tubercle, posterior angles rounded; elytra with five punctured striae, interstices flat. L. 1—1½ l. Common.

A. paniceum, Lin.

Xestobium.

Blackish-red-brown, finely and extremely closely punctured; upper-side sprinkled with little spots of yellow hairs. L. 2¾—4 l. Common.   *X. tesselatum*, Fab.

Ernobius.

A. Thorax uneven.

Reddish-yellow-brown above, pitchy beneath. Thorax much broader than long, anterior angles almost right angles, scarcely blunted; antennae with third joint not longer than second, fifth much longer than sixth, and eighth almost transverse. L. 1½ l. Rare.   *E. abietis*, Fab.

B. Thorax even.

a. Body black.

Tarsi reddish. Thorax broader than long, anterior angles obtuse and strongly rounded; antennae with third joint distinctly longer than second, fifth much longer than sixth, and eighth distinctly transverse. L. 2 l. Rare.

E. nigrinus, Sturm.

b. Body reddish-brown.

Antennae, apex of elytra and legs pale, eyes black.
Thorax much broader than long, anterior angles somewhat obtuse, slightly rounded; antennae with third joint
much longer than second, fifth distinctly longer than sixth,
and eighth considerably longer than broad. L. 2 l. Common. *E. mollis*, Lin.

Ptilinus.

Black; elytra brown; antennae and legs reddish-brown.
Anterior angles of thorax somewhat obtuse and slightly
rounded; elytra confusedly punctured, without raised
lines. L. 2—2½ l. Common. *P. pectinicornis*, Lin.

Ochina.

Brown; antennae and legs lighter; elytra with two
transverse bands of gray hairs. Sides of thorax slightly
rounded and feebly bordered, anterior angles almost right
angles. L. 1—1¼ l. Moderately common.
O. hederae, Müll.

Xyletinus.

Oblong. Black, with silky reflection; antennae blackish
brown; tibiae and tarsi reddish-brown. Sides of thorax
slightly rounded, anterior angles right angles; elytra with
fine punctured striae. L. 1½ l. Scarce *X. ater*, Panz.

Lasioderma.

Brownish-red; antennae and legs lighter; eyes black.
Thorax double as broad as long, slightly rounded at sides,
much narrowed in front; pubescence on elytra arranged
somewhat in lines; third joint of antennae scarcely so
long as second; second joint of tarsi half as long as first
and not longer than third. L. 1 l. Rare.
. *L. serricornis*, Fab.

Dorcatoma.

A. Elytra rather closely, not very strongly punctured,
pubescence slightly recumbent, both longitudinally and
transversely.

Black, shiny; antennae and legs rust-red. Forehead
gradually somewhat narrowed in front, moderately convex;

ventral abdominal segments free.   L. 1 l.   Not common.
D. chrysomelina, Sturm.

B. Elytra closely punctured in wrinkles, pubescence recumbent longitudinally only.
Black, shiny; antennae and legs reddish.   Forehead broad, rather convex; second to fourth ventral abdominal segments more or less soldered in middle.   L. 1 l.   Rather common.   *D. flavicornis*, Fab.

Coenocara.

Black, shiny; antennae and legs brown.   Elytra confusedly punctured, with three impunctate striae at sides, pubescence not arranged in rows.   L. $\frac{3}{4}$—1 l.   Not common.   *C. bovistae*, Hoff.

Anitys.

Rust-red, shiny; antennae and legs lighter.   Elytra with traces of a sutural stria behind and slight traces of others on disc; posterior tibiae shorter than femora and trochanters together; tarsi very short and thick.   L. 1 l.   Scarce.
A. rubens, Hoff.

Sphindus.

Almost cylindrical.   Black-brown; shoulders of elytra, antennae and legs reddish-yellow-brown.   Thorax finely punctured; elytra with punctured striae, pubescence short, arranged in rows.   L. 1 l.   Scarce.   *S. dubius*, Gyll.

CONIPORIDAE.

Last three joints of antennae forming an elongate club; clypeus separate from forehead; inner lobe of maxillae with a horny hook at apex; last joint of palpi awllike.
Conipora, Th.

Conipora.

Roundish, convex.   Black or brown; antennae and legs reddish-yellow-brown.   Thorax very finely punctured; elytra much broader than thorax, deeply punctured.   L. $\frac{1}{2}$ l.   Not common.   *C. orbiculata*, Gyll.

BOSTRYCHIDAE.

A. Second joint of tarsi elongate.

Labrum rather small; last joint of maxillary palpi about equal to third; paraglossae prominent; antennae ten-jointed, club rarely as long as funiculus, not toothed, more or less loose; apex of elytra either rounded or obliquely truncate; anterior tibiae generally simple; tarsi long and narrow, second joint as long as third and fourth together.

Bostrychus, Geoffr.

B. Second joint of tarsi short.

a. Club of antennae shorter than funiculus, not distinctly toothed.

Labrum large; last joint of maxillary palpi scarcely longer than third; paraglossae long, rather narrow; antennae ten-jointed; elytra rounded and entire at apex; anterior tibiae toothed; tarsi with first four joints about equal in length. *Dinoderus*, Redt.

b. Club of antennae about equal in length to funiculus, strongly sawlike.

Labrum large; last joint of maxillary palpi almost as long as second and third together; paraglossae rather short and obtuse; antennae ten-jointed; elytra simply retuse at apex; anterior tibiae toothed; tarsi with first joint very small, next three about equal in length. *Rhizopertha*, Steph

Bostrychus.

Black; elytra and abdomen red. Thorax not emarginate in front, closely granulate, the granules larger in front and at sides; elytra deeply, confusedly punctured. L. 3—5½ l. Very rare. *B. capucinus*, Lin.

Dinoderus.

Pitch-black, pubescence brownish. Thorax and elytra closely granulate, the granules forming little teeth in front and at sides of former, and rather indistinct rows on middle of latter. L. 2½ l. Rare. *D. substriatus*, Payk.

Rhizopertha.

Pitchy-red; antennae dull-red. Thorax convex, rough in front; elytra bare, shiny, with punctured striae. L. 1¾ l. Common. *R. pusilla*, Fab.

LYCTIDAE.

Clypeus separate from forehead; last joint of maxillary

G 2

palpi oval and of labial palpi conical, both pointed at apex ; eyes rather large and prominent.   *Lyctus*, Fab.

Lyctus.

A. Head granulate.

Pitch-brown or brown ; antennae and legs rust-red. Head and thorax granulate, latter scarcely narrowed behind, with a broad central furrow, sides straight, finely notched, posterior angles right angles ; elytra with striae, interstices finely punctured, pubescence in rows. L. 2— 2½ l. Common.   *L. canaliculatus*, Fab.

B. Head delicately punctured.

Elongate. Brown, glabrous ; head pitchy, eyes black ; thorax pitch-brown ; elytra bright chestnut-brown ; antennae and legs reddish-pitchy. Head delicately punctured ; thorax coarsely punctured, with a broad, shallow, central furrow, elytra with fine striae and very faintly punctured. L. 2½ l. Rare.   *L. brunneus*, Steph.

CISSIDAE.

A. Antennae ten-jointed.

    a. Tibiae scarcely dilated at apex, without spines on outer-side.

Last joint of maxillary palpi long oval ; head bordered in front.   *Cis*, Latr.

    b. Tibiae dilated at apex, with spines on outer-side.

Second joint of antennae much longer than third ; head not bordered, toothed in middle of front ; last joint of maxillary palpi somewhat oblong.   *Rhopalodontus*, Mell.

B. Antennae nine-jointed.

Second joint of antennae much smaller than first ; tibiae with some spines toward apex.   *Ennearthron*, Mell.

C. Antennae eight-jointed.

Second joint of antennae not very much shorter than first ; head and thorax of male without tubercles ; tibiae with spines on outer-side.   *Octotemnus*, Mell.

Cis.

A. Thorax uneven, indistinctly ridged in middle.

a. Thorax not bordered at base.

I. Elytra rugulose, with large punctures.

Pitch-black, brown or yellow-brown ; antennae and legs rust-red. Closely covered with extremely short hairs ; thorax rather uneven, anterior margin raised and sinuate, side margin broadly separate, with a fine raised line near it, base not bordered ; elytra very finely and closely punctured, with an admixture of larger punctures, forming more or less distinct rows toward base. L. 1—1¾ l. Common.

C. boleti, Scop.

II. Elytra rugulose, without large punctures.

Nearly cylindrical. Pitch-black ; closely covered with short, shiny pubescence ; antennae and legs pale-brown or rust-red. Thorax transverse, anterior margin raised and sinuate, sides broadly bordered ; base not bordered. L. 1¾ l. Rare. *C. rugulosus,* Mell.

b. Thorax bordered at base.

Nearly cylindrical. Pitch-black, strewn with golden-yellow scale-like hairs ; antennae and legs pale-brown or rust-red. Thorax transverse, very uneven, ridged, anterior margin raised, lateral border narrower than in *C. rugulosus* and more strongly bristly ; elytra rugulose, without large punctures. L. 1¾ l. Rather common.

C. villosulus, Marsh.

B. Thorax even.

a. Elytra very finely punctured in wrinkles, without rows of punctures.

Pitch-brown ; antennae and legs lighter ; pubescence very short, shiny, yellow. Thorax sinuate in front, finely bordered at sides and behind. L. 1 l. Common.

C. micans, Herbst.

b. Elytra punctured in wrinkles, with rows of punctures.

I. Thorax bordered behind.

Pitch-black or pitch-brown ; antennae and legs yellow-brown, evenly and closely covered with very short, stiff hairs. Thorax slightly sinuate in front, finely bordered at sides and behind ; rows of punctures on elytra feeble but distinct in front. L. 1 l. Not common.

C. hispidus, Payk.

II. Thorax not bordered behind.

Elongate, rather flat.   Brown ; pubescence very short, shiny, arranged in rows.   Thorax straight in front, bordered at sides, very closely and finely punctured ; elytra as broad as thorax, with tolerably regular rows of punctures.   L. ¾ l.   Rare.   *C. elongatulus*, Gyll.

c. Elytra finely punctured (not in wrinkles).

I. Anterior tibiae produced into a distinct tooth on outer side of apex.

1. Posterior angles of thorax not rounded ; prosternum without ridge before coxae.

Black ; pubescence very short and fine.   Anterior angles of thorax prominent, posterior angles rounded ; elytra hairy, very closely punctured.   L. ¾—1 l.   Common.

C. bidentatus, Ol.

2. Posterior angles of thorax rounded ; prosternum with a short ridge before coxae.

A A. Anterior angles of thorax slightly prominent.

Pitch-brown, shiny ; antennae (except club) and legs red-yellow.   Posterior angles of thorax rounded, sides and base bordered ; elytra unevenly punctured.   Clypeus of male with two indistinct teeth.   L. ⅘ l.   Not common.

C. nitidus, Herbst.

B B. Anterior angles of thorax not prominent.

Similar to *C. nitidus* but with anterior angles of thorax nearly right angles, not produced, forehead with a depression, and clypeus of male with a rather acute tooth on each side. L. ¾ l.   Rare.   *C. Jacquemarti*, Mel.

II. Anterior tibiae not or only slightly produced on outer-side of apex,

Anterior angles of thorax not prominent.

1. Posterior angles of thorax almost right angles.

A A. Body shiny.

a a. Body pitch-black; pubescence very short.

Elongate.   Mouth, antennae and legs light yellow-brown. Thorax bordered broadly at sides, narrowly at base.   L. ¾ —1 l.   Rather common.   *C. alni*, Gyll.

b b. Body brown-yellow; pubescence long.

Elongate, almost cylindrical. Antennae and legs light yellow-brown. Thorax bordered rather broadly at sides, narrowly at base, with a transverse impression before latter, Head of male with two indistinct tubercles. L. 1 l. (? Brit.)

C. punctulatus, Gyll.

B B. Body dull.

Somewhat oblong. Black, dull, strewn with short, shiny pubescence; antennae and legs pale rust-red. Thorax with sides broadly and base narrowly bordered; elytra very closely, finely and shallowly punctured. Head of male with two tubercles. L. 1 l. Moderately common.

C. pygmaeus, Marsh.

2. Posterior angles of thorax obtuse or rounded.

A A. Body without pubescence.

Rather short, convex, shiny, scarcely pubescent. Rust-red; antennae and legs pale red-yellow. Mandibles scarcely reaching beyond labrum; thorax nearly impunctate, sides and base bordered; elytra diffusely and deeply punctured in rows. L. $\frac{2}{3}$ l. Not uncommon.

C. lineato-cribratus, Mel.

B B. Body pubescent.

a a. Body somewhat oval.

Convex, somewhat oval. Brown-yellow, rather shiny; pubescence short. Thorax bordered broadly at sides, narrowly at base; elytra with an indistinct sutural furrow; punctuation distinct. Head of male with two tubercles. L. $\frac{4}{5}$ l. Moderately common.   *C. festivus,* Panz.

b b. Body nearly cylindrical.

A a. Body yellow-brown; pubescence not very short.

Nearly cylindrical, rather convex. Yellow-brown, pubescence longer than in *C. festivus.* Thorax dull, very closely punctured, narrow, contracted in front, sides and base bordered; elytra irregularly and closely punctured. Head of male with two tubercles. L. $\frac{3}{4}$ l. Rare.

C. vestitus, Mel.

B b. Body brown; pubescence very short.

Elongate, parallel-sided. Brown, with very short, shiny pubescence. Sides and base of thorax bordered; elytra very closely punctured. Head of male with two tubercles. L. ¾ l. Not uncommon. *C. fuscatus*, Mel.

Rhopalodontus.

A. Length ⅘ line.

Oblong, convex. Pitch-black; pubescence rather long and thin. Thorax short, rounded at sides and angles, bordered at base ; elytra punctured in wrinkles, with scattered larger punctures. Moderately common.

R. perforatus, Gyll.

B. Length ⅖—½ line.

Black or pitch-black, shiny ; antennae yellow-brown, club blackish ; legs red-brown, femora sometimes darker ; rather closely (especially on elytra) covered with very short, upright bristles. Thorax much broader than long, very finely bordered at sides and base, with angles rounded, base tolerably straight, anterior margin somewhat produced, not sinuate ; elytra as broad as thorax, scarcely half as long again as broad. Forehead of male with two straight, little horns in middle of front and a blunt tubercle above insertion of antennae, that of female simple. Not uncommon.

R. fronticornis, Panz.

Ennearthron.

A. Body pitch-black or pitch-brown.

Covered with stiff, upright, whitish hairs. Thorax bi-sinuate in front, anterior angles produced, tolerably rounded at sides, narrowly bordered, very closely punctured ; elytra more strongly punctured than thorax, pubescence mostly in rows, especially close behind. Head of male with two straight black horns in front, that of female simple. L. 1 l. Rather common. *E. affine*, Gyll.

B. Body reddish-brown or yellow-brown.

Convex. Not very closely covered with very short, stiff hairs ; punctuation fine and scattered, that of elytra rather stronger. Front of forehead with two small black horns in male, slightly raised in female ; front of thorax strongly produced, in male deeply emarginate in middle (forming two teeth), in female rounded. L. 1 l. Rather common.

E. cornutum, Gyll.

Octotemnus.

Chestnut-brown, without pubescence. Thorax very finely punctured; elytra closely punctured, here and there somewhat in wrinkles. L. $\frac{1}{2}$—$\frac{3}{4}$ l. Common.

O. glabriculus, Gyll.

RHYNCHOPHORA.

A. Head not produced into a rostrum, or only slightly so; in latter case outer margin of tibiae toothed.

Mentum prominent, generally received at base into an emargination of submentum; maxillae nearly always with one lobe; palpi short; antennae inserted almost always between eyes and mandibles or on sides of rostrum, short, elbowed and clubbed, with from three to twelve joints; eyes generally large and transverse; pronotum rarely separate from sides of prothorax; tibiae flattened, nearly always toothed on outer margin; tarsi not spongy beneath, thread-like, third joint entire or bilobed but never much dilated; abdomen with five segments.   *Scolytidae.*

B. Head produced into a longer or shorter rostrum; outer margin of tibiae not toothed.

a. Labrum absent.

Mentum placed in an emargination of submentum or carried by a stalk rising from bottom of latter; maxillae generally with one lobe; palpi very short; antennae inserted on rostrum, generally elbowed, and nearly always clubbed, with from eight to twelve joints; pronotum not separate from sides of prothorax; tarsi generally spongy beneath, third joint usually bilobed; abdomen with five, rarely six segments, third and fourth usually shorter than the others.   *Curculionidae.*

b. Labrum distinct, rounded and fringed in front.

Sub-mentum provided with a large and broad bilobed stalk, enclosing the mentum and ligula between its lobes; maxillae with two lobes mandibles more or less prominent, flattened, dilated and toothed at base, curved and acute at apex; antennae inserted on rostrum, not elbowed, generally clubbed, sometimes threadlike, with eleven joints; thorax with a transverse ridge near or at base; anterior coxae placed not far apart; apex of tibiae truncate, not hooked, second tarsal joint bilobed, third very small, received between lobes of second; claws free, toothed beneath; abdomen with five nearly equal segments; pygidium not covered. *Anthribidae.*

SCOLYTIDAE.

A. First tarsal joint as long as the other three together.

Head not sunk in thorax, vertical or oblique; ligula represented by a ridge on mentum; maxillary palpi with four, labial with from one to three joints; funiculus of antennae with four joints, club solid; thorax more or less cylindrical, sides with cavities for reception of anterior femora; anterior coxae very large, ovoid, oblique; tarsi long, third joint entire; episterna of mesothorax very large, rounded in front. *Platypides.*

B. First tarsal joint much shorter than the other three together.

Head sunk in thorax; ligula free, at least partly, and reaching beyond mentum in front; all palpi with three joints; labrum always indistinct; thorax hoodlike, sides without cavities for reception of anterior femora; anterior coxae more or less prominent, scarcely ever oblique; third tarsal joint entire or bilobed; episterna of mesothorax very rarely large. *Scolytides.*

PLATYPIDES.

Labial palpi two-jointed; maxillae with one lobe; maxillary palpi membranous, very large, the joints embedded in one another; labrum very short, inconspicuous; eyes oval or oblong oval; elytra elongate, cylindrical, covering pygidium, apex sometimes produced; anterior coxae placed close together. *Platypus,* Herbst.

Platypus.

Elongate, cylindrical. Dark brown, disc of elytra usually lighter in female; antennae and legs red-brown. Elytra with punctured striae. L. 2—2½ l. Moderately common. *P. cylindrus,* Fab.

SCOLYTIDES.

A. Abdomen turned upward from second segment onward.

Head with scarcely any trace of rostrum, more or less visible from above ; antennae with seven joints to funiculus, club compact ; eyes finely granulate, narrow ; pronotum separate from sides of prothorax ; elytra without apical declivity ; second abdominal segment as long as the next two together ; third tarsal joint bilobed. *Scolytus*, Geoffr.

B. Abdomen normal.

a. Head globular, generally invisible from above.

Eyes finely granulate ; pronotum not separate from sides of prothorax ; anterior coxae placed close together ; third tarsal joint simple ; metasternum more or less long, its episterna narrow.

I. Funiculus of antennae with five joints.

1. Apical declivity of elytra more or less hollowed out.

Mentum at least four times as long as broad, narrowed behind middle, again dilated toward apex, anterior margin cut very obliquely on each side, forming an acute angle ; ligula commencing near middle of mentum and there as broad as it, strongly narrowed in front, apex rounded ; club of antennae rather small, separation of joints not very distinct, the sutures more or less curved ; eyes narrow, sinuate in front ; second to fourth abdominal segments about equal in size. *Tomicus*, Latr.

2. Apical declivity of elytra rarely slightly impressed.

A A. First joint of labial palpi not much longer than second.

a a. Maxillary lobes fringed with a few stiff, pointed, tolerably straight bristles.

Ligula almost three times as long as broad, somewhat narrowed behind middle ; first and second joints of labial palpi nearly equal, third much narrower, slightly conical, not shorter than second ; club of antennae with first joint nearly round, second and third crescent-shaped ; second tarsal joint a little shorter than first and third. *Xylocleptes*, Ferr.

b b. Maxillary lobes fringed with almost straight, broad, thornlike teeth.

A a. Mentum elongate triangular, very narrow at base.

Ligula commencing at base of mentum narrow, parallel-sided, tolerably acute in front ; club of antennae oval, very distinctly four-jointed, sutures straight.

Pityophthorus, Eich.

B b. Mentum rather long, slightly dilated in front.

Ligula commencing near base of mentum, narrow, nearly parallel-sided, rounded in front ; similar in other respects to *Xyleborus*.

Dryocaetes, Eich.

B B. First joint of maxillary palpi very large, inflated.

Ligula commencing near anterior third part of mentum, narrow, oblong oval, apex acute ; club of antennae ovoid, distinctly jointed, sutures straight.

Xyleborus, Eich.

II. Funiculus of antennae with four joints.

1. Eyes narrow, entire.

Mentum elongate, gradually dilated and nearly rounded in front; joints of labial palpi gradually smaller ; club of antennae compressed, oval, shewing fine straight or curved sutures; scutellum very small ; apical declivity of elytra rounded ; body covered with fine hairs and generally with some scales.

Cryphalus, Er.

2. Eyes oblong-ovate, deeply emarginate.

Mentum very elongate, gradually dilated in front, anterior margin angular in middle ; first two joints of labial palpi equal, third very small ; club of antennae compact, oblong oval; scutellum moderately large, triangular; apical declivity of elytra rounded ; second abdominal segment a little longer than each of the two following ones.

Trypodendron, Steph.

III. Funiculus of antennae with three joints.

Mentum elongate, parallel-sided; first two joints of labial palpi equal, third very small ; funiculus of antennae with first joint very large, second and third transverse, equal, club very large, compressed, oval, shewing traces of sutures ; apical declivity of elytra rounded. *Hypothenemus*, Westw.

b. Head not globular, visible from above.

Eyes finely granulate; pronotum not separate from sides of prothorax ; metasternum more or less long, its episterna narrow.

I. Club of antennae solid ; third tarsal joint entire.

Mentum elongate, gradually dilated and cut obliquely on each side in front; ligula commencing near apex of mentum ; first joint of labial palpi larger than second ; funiculus of antennae with five joints, club solid, short oval ; eyes nearly divided ; first three tarsal joints equal, third entire, not broader than second ; intermediate abdominal segments very short, equal. *Polygraphus*, Er.

II. Club of antennae distinctly articulate ; third tarsal joint emarginate or bilobed.

1. Club of antennae elongate.

A A. Funiculus of antennae with five joints, club formed of three joints, loosely united.

Mentum elongate, gradually dilated and angular in middle of front; ligula commencing before middle of mentum ; labial palpi longer than maxillary, first joint longer than second ; first and second tarsal joints nearly equal, third a little dilated, bilobed. *Phlocophthorus*, Woll.

B B. Funiculus of antennae with seven joints, club formed of four joints, closely united.

Mentum broad, heartshaped, very narrow at base, angular in middle of front; ligula commencing at some distance from base of mentum ; first joint of labial palpi longer than second and third together ; anterior coxae placed slightly apart; first tarsal joint a little longer than second, third dilated, bilobed ; second abdominal segment almost as long as next two together ; metathoracic episterna tolerably broad. *Hylesinus*, Fab.

C C. Funiculus of antennae with six joints; club formed of four joints, closely united.

Mentum rotundate ovate at base ; third tarsal joint distinctly bilobed. *Cissophagus*, Chap.

2. Club of antennae oval or nearly globular, formed of four joints, closely united.

Anterior coxae close together.

A A. Funiculus of antennae with five joints.

Mentum elongate, gradually dilated in front, middle of anterior margin slightly angular ; ligula commencing in middle of mentum ; labial palpi short, first joint a little longer than second ; funiculus of antennae with only its first joint elongate ; first tarsal joint much shorter than second, third not dilated, emarginate ; second abdominal segment almost as long as next two together.

Carphoborus, Eich.

B B. Funiculus of antennae with six joints.

Mentum scarcely longer than broad in front, base much narrowed, front broadly rounded ; ligula commencing at base of mentum ; first joint of labial palpi large, the other two very small ; third tarsal joint not broader than second, emarginate; intermediate abdominal segments equal, short.

Hylurgus, Latr.

C C. Funiculus of antennae with seven joints.

Mentum rather long, gradually slightly dilated in front, middle of anterior margin somewhat angular ; ligula commencing at base of mentum ; first joint of labial palpi large, the other two very small ; third tarsal joint heart-shaped or bilobed ; intermediate abdominal segments equal short.

Hylastes, Er.

Scolytus.

A. Thorax diffusely and not strongly punctured.

a. Male with prominences on third and fourth ventral segments of abdomen.

I. Elytra black.

Black, shiny ; apex of femora narrowly, and of tibiae broadly pitchy-red ; tarsi red-yellow. Forehead of male closely covered with long erect hairs, that of female scantily covered with long, depressed hairs, ridged ; thorax a little longer than broad, coarsely and deeply punctured at sides and front, finely and more diffusely on disc and behind ; elytra parallel-sided, with strong punctured striae, interstices flat, with a somewhat irregular row of very fine punctures on each ; second ventral segment of abdomen nearly vertical, very diffusely and rather indistinctly punctured, fifth segment with a broad, deep impression, punctuation coarser and slightly closer. L. 2½ l. Not common ; in Birch. S. Ratzeburgi, Jans.

II. Elytra brown or red-brown.

Black; elytra brown or red-brown; antennae and legs reddish-yellow-brown. Forehead of male with short pubescence, that of female not ridged; clypeus less deeply emarginate than in *S. Ratzeburgi*, thorax broader, elytra less parallel-sided, abdomen with a minute tooth at anterior margin of third and fourth segments in both sexes and punctured (although finely) more deeply and closely, especially on apical segment. L. 2¼ l. Common; in Elm.

S. destructor, Ol.

b. Abdominal ventral segments simple.

Black, shiny; anterior and posterior margins of thorax and the elytra brown; antennae and legs red-brown. Thorax almost as long as broad, narrowed in front; elytra with fine punctured striae, interstices with a still finer row of punctures. L. 1¾—2 l. Not common. *S. pruni*, Ratz.

B. Thorax rather closely and strongly punctured.

a. Abdominal ventral segments of male simple.

I. Interstices on elytra rather more finely punctured than the striae; elytra unicolorous.

Black; shiny; antennae and legs rust-red; femora and elytra pitch-brown, latter with very close rows of punctures, with short, erect yellow bristles at sides and toward apex; under-side with close, gray pubescence. Thorax almost broader than long, rather strongly and closely punctured, more feebly in middle, coarsely and almost in wrinkles at sides; suture depressed close behind scutellum only, without trace of sutural furrow. L. 1¼ l. Not uncommon. *S. intricatus*, Ratz.

II. Punctures on interstices of elytra as large and deep as those in striae; apex of elytra reddish-brown.

Black, shiny; apex of elytra, antennae and legs reddish-brown. Thorax extremely closely covered with deep, oblong punctures, confluent into wrinkles in front and at sides. L. ¾—1 l. Rather common. *S. rugulosus*, Ratz.

b. Second ventral segment of abdomen in male with a large, horizontal, prominent tooth.

Black; elytra brown, apex (rarely entirely) red-brown, antennae yellow-brown; legs red. Elytra with very close, nearly equally strong punctured striae, apex confusedly punctured.   L. 1—1½ l.   Rather common.

S. *multistriatus*, Marsh.

Tomicus.

A. First joint of club of antennae angularly produced above ; prosternal process shorter.

a. Apical impression of elytra with six teeth on each side, the three upper ones being small and the fourth largest.

Black, with brown elytra, or entirely brown or yellow-brown.   Similar to *T. typographus* but with hinder part of thorax strewn with deeper punctures (except smooth central line) and elytra with stronger punctured striae. L. 3—3¾ l.   Rare.   *T. stenographus*, Duft.

b. Apical impression of elytra with four teeth on each side, the uppermost one often indistinct and the third largest.

Black, with brown elytra, or entirely brown or yellow-brown.   Front half of thorax closely granulate, hinder part finely and diffusely punctured ; scutellum small, smooth, even ; elytra with fine punctured striae, punctures feebler behind.   L. 2—2½ l.   Rare.   *T. typographus*, Lin.

c. Apical impression of elytra with three teeth on each side, the lowest one largest.

Lighter or darker brown ; antennae and legs yellow-brown ; moderately shiny ; with long, yellow-gray hairs. Thorax rather closely punctured, granulate in front, without smooth central line behind ; elytra flattened in a circle at apex, margin of impression nearest suture scarcely raised, its lateral margin with three teeth, of which the uppermost is only a little prominence and the lowest one (placed about middle of margin) is a rather long, pointed tooth.   Elytra shorter than in preceding species, with striae shallower, tibiae less dilated and first joint of club of antennae less produced.   L. 1½—1¾ l.   Rather common.

T. acuminatus, Gyll.

B. First joint of club of antennae not angularly produced above ; prosternal process longer.

a. Club of antennae rounded at apex, sutures between its joints straight.

Lighter or darker brown or yellow-brown. Thorax closely granulate in front, finely and diffusely punctured behind; interstices on elytra with a row of isolated punctures, apical impression with from three to six small, straight teeth on each side and a little tooth inward from the second and third teeth. L. 1½—1¾ l. Common.

T. laricis, Fab.

b. Club of antennae nearly truncate at apex, sutures between its joints curved.

Very similar to *T. laricis,* but with hairs white, punctured striae on elytra less strong, apical impression not so extensive, more obsoletely punctured and less strongly toothed at sides, club of antennae nearly truncate at apex, sutures between its joints curved. L. 1½ l. Rare,

T. nigritus, Gyll.

Xylocleptes.

Brown, shiny. Thorax with straight sides, distinctly broader in front, fore part of disc closely granulate, hinder part (especially in female) closely punctured, except smooth central line; elytra with fine and rather close punctured striae, their apex in male impressed, with a large erect tooth, in female flattened and uneven on account of the raised suture and some rows of granules. L. 1¼—1¾ l. Common.

X. bispinus, Duft.

Pityophthorus.

A. Prosternum with distinct process; base of thorax evidently bordered.

Pitchy-red, shiny; antennae and legs red-yellow. Thorax diffusely and finely punctured behind, central line impunctate; elytra with fine punctured striae, interstices almost impunctate, with indistinct transverse scratches. L. 1 l. Moderately common.

P. micrographus, Gyll.

B. Prosternum without process; base of thorax not bordered in middle.

a. Thorax diffusely and finely punctured behind.

Either entirely reddish-yellow-brown or with thorax and base of elytra dark brown, very shiny. Thorax closely granulate in front, diffusely and finely punctured behind, with smooth central line; elytra with extremely fine punctured striae, interstices smooth, strongly and broadly impressed along suture, with three strong, toothlike prominences (placed longitudinally, parallel with suture) on each side. L. $\frac{3}{4}$—$\frac{4}{5}$ l.   Scarce.   *P. chalcographus*, Lin.

b. Thorax punctured rather closely and strongly behind.

 I. Elytra of male with upper tooth at apical impression large and lower tooth small; elytra of female with a tubercle above furrow at apex.

Usually lighter or darker brown, often with head and thorax black. Thorax narrowed in front, fore part closely granulate, hinder part rather closely punctured, with smooth and somewhat raised central line; elytra with fine punctured striae, their apex in male impressed, with a large, hooked tooth at commencement of impression and above this usually a small prominence, in female with only suture raised and a narrow furrow near it on each side.   L. 1 l. Common.   *P. bidens*, Fab.

 II. Elytra of male with upper tooth at apical impression not large and lower tooth not small; elytra of female with a tubercle placed at end of furrow at apex.

Black, shiny; antennae rust-red; legs brown.   Thorax strongly punctured behind, with a narrow, somewhat raised central line and a round lateral spot impunctate.   Elytra with fewer and shorter hairs than *P. bidens*, disc with finer rows of punctures, the apical impression in male obtusely bordered beneath, not fringed with hair, tooth on upper part feebler, lower one more distinct, tubercle at apex of elytra of female placed lower.   L. 1 l.   Rare.

            P. quadridens, Nord.

Dryocaetes.

A. Thorax evenly punctured throughout.

 a. Sides of thorax somewhat rounded; length 2 lines.

Oblong, shiny.   Brown.   Similar to *D. villosus* but with thorax shorter and broader, disc less closely punctured, apex of elytra less abruptly retuse, sutural stria less defined, teeth on tibiae feebler and placed more closely, first joint of club of antennae longer.   Rare.   *D. autographus*, Ratz.

 b. Sides of thorax nearly parallel; length 1½ lines.

Oblong, shiny.   Lighter or darker brown, covered with

long, pale hairs. Thorax very closely punctured, elytra with strong punctured striae and a row of fine punctures on each interstice, apex somewhat retuse, sutural stria deep behind. Common. *D. villosus*, Fab.

B. Thorax granulate throughout.

Black; elytra black-brown or brown, shiny; antennae and legs yellow-brown. Thorax dull, closely covered with transverse, scalelike granules; elytra with rows of strong punctures, sutural stria deep, interstices with a row of much more diffusely placed fine punctures. Narrower than *D. bicolor* and less hairy, thorax without obsolete transverse depression behind middle and elytra much less distinctly flattened at apex. L. 1—1¼ l. Rare. *D. alni*, Georg.

C. Thorax granulate in front, punctured behind.

a. Hinder part of thorax diffusely punctured, with impunctate central line; body brown-red.

Narrow, cylindrical, shiny. Brown-red, thorax lighter in front; antennae and legs red-yellow; pubescence gray. Thorax half as long again as broad, closely granulate in front, diffusely and finely punctured behind, middle impunctate, sides straight or scarcely sinuate; elytra obliquely flattened at apex, with rows of strong punctures, impressed near suture behind. L. ¾—1 l. Rare. *D. coryli*, Per.

b. Hinder part of thorax closely punctured, without impunctate central line; body pitch-black or brown.

Pitch-black or brown, with long whitish-gray hairs; antennae and legs pale yellow-brown. Front half of thorax strongly but not closely granulate, hinder part closely punctured; apical declivity of elytra forming a round, finely punctured surface (almost sharply defined at sides), with a narrow, shallow furrow near the somewhat raised suture. L. ¾ l. Scarce. *D. bicolor*, Herbst.

Xyleborus.

A. Elytra together as broad or nearly as broad as long.

Short. Pitch-black; antennae and legs reddish-yellow-brown. Front of thorax closely granulate, hinder part finely punctured; elytra with punctured striae, interstices broad, with a row of much finer punctures, in male almost globular, broader than thorax, together as broad as long, in

н 2

female short cylindrical, about one quarter longer than together broad. L. male ⁴⁄₅ l., female 1¼ l. Rare.

X. dispar, Fab.

B. Elytra double as long as together broad.

a. Hinder part of thorax finely and diffusely punctured.

Pitch-brown or reddish-brown. Thorax longer than broad, front convex and closely granulate ; elytra with punctured striae, interstices with a row of fine punctures, flattened apical declivity with two or three rows of small granules. L. 1—1¼ l. Not uncommon.

X. dryographus, Er.

b. Hinder part of thorax impunctate.

Very similar to *X. dryographus*, but slightly smaller; apical declivity of elytra more flattened, with many more distinct and pointed granules, continuing first, third and (partly) fourth striae, with a broad furrow in place of second row. L. 1¼ l. Not common. *X. Saxeseni*, Ratz.

Cryphalus.

A. Body more or less short cylindrical.

a. Anterior margin of thorax with four small teeth.

I. Rows of punctures on elytra distinct throughout.

1. Thorax almost regularly, coarsely granulate in front, punctures of striae on elytra scarcely coarser than those of interstices ; length 1 line.

Black, shiny; funiculus of antennae and part of legs lighter. Very similar to *C. binodulus*, but elytra throughout their breadth with distinct rows of punctures, the first two of which appear impressed on apical declivity. Rare.

C. granulatus, Ratz.

2. Thorax with three concentric rows of granules ; elytra with interstices punctured more finely than striae ; length ½ line.

Brown or yellow-brown, dull, with fine hairs ; antennae and legs and usually also elytra lighter. Thorax rounded, with concentric rows of small prominences on front half; elytra with feeble punctured striae. Scarce. *C. tiliae*, Fab.

II. Rows of punctures on elytra distinct on outer part only.

Black, shiny; scantily covered with scalelike hairs and other short hairs; antennae and legs dirty yellow. Anterior margin of thorax with four prominences in middle and with concentric rows of granules, united in places almost into sharp lines; elytra with feeble punctured striae, effaced toward suture, and two prominences on apical declivity. L. ½ l. Not uncommon. *C. binodulus*, Ratz.

b. Anterior margin of thorax without teeth.

I. Thorax with five or six crowded rows of granules in front; rows of punctures on elytra very fine.

Very similar to *C. abietis* but generally rather larger; granules on thorax forming a tolerably broad rhombus and arranged in five or six regular, crowded rows; elytra with tolerably distinct punctured striae; club of antennae pointed at apex. L. ⅔ l. Rare. *C. piceae*, Ratz.

II. Thorax confusedly granulate; rows of punctures on elytra not very fine and placed in somewhat impressed striae.

Compressed and very convex. Black-brown; antennae and legs reddish-brown. Thorax almost globular, very finely punctured at sides and behind, front part with diffuse granules, standing in rows here and there only; elytra rather more than double as long as thorax, with distinct punctured striae and extremely finely punctured interstices, covered throughout with extremely fine, close-lying scalelike hairs (with a reddish-brown-gray reflection) and also very diffusely with short, erect hairs, effaced behind; club of antennae rounded at apex. L. ½—⅔ l. Not uncommon. *C. abietis*, Ratz.

B. Body elongate cylindrical.

Black, shiny; elytra, antennae and legs dirty yellowish. Front part of thorax uneven and (when viewed from front) transversely wrinkled. L. ½ l. Scarce. *C. fagi*, Nord.

Trypodendron.

A. Club of antennae produced at inner-side of apex.

a. Elytra almost double as long as together broad.

Cylindrical. Black; sides of thorax rarely brownish; elytra yellow-brown, with suture and side margin blackish; antennae yellow; legs black or brown, tarsi yellow. Club of antennae dilated toward apex, which is rounded and produced within into a little tooth ; front half of thorax rather strongly granulate, the granules crowded in middle into a short raised transverse line ; elytra nearly twice as long as together broad, apex deeply furrowed near suture, rows of punctures rather feebler than in *T. lineatum*. L. 1½—1¾ l. Not common.   *T. domesticum*, Lin.

b. Elytra half as long again as together broad.

Short cylindrical. Black ; antennae, legs, part of thorax and the elytra brown-yellow, latter with suture, outer margin and a streak on disc black. Larger than *T. lineatum*, with club of antennae larger and more widened toward front, rugosities of thorax coarser and pubescence closer and longer, punctures of striae on elytra deeper, but not so sharp-edged, lineation blacker and more defined. L. 1¾ l. Scarce.   *T. quercus*, Eich.

B. Club of antennae rounded at apex [*Xyloteres*, Er.]

Short cylindrical. Black ; more or less of thorax and the elytra yellow-brown, latter with suture, side margin and often a streak on disc blackish. Thorax almost globular, anterior margin not raised, disc with fine, scalelike, raised transverse wrinkles and punctures ; elytra half as long again as broad, with distinct large rows of punctures, apex feebly furrowed near suture. L. 1¼—1½ l. Rare.
     T. lineatum, Er.

Hypothenemus.

Cylindrical. Pitch-black, covered with fine hairs; thorax brownish-yellow, convex and tuberculate in front; antennae and legs yellowish. Elytra more than twice as long as thorax, with punctured striae, apex rounded, entire. L. ⅘ l. Extremely rare.   *H. eruditus*, Westw.

Polygraphus.

Black, brown or yellow-brown, with short, scale like hairs ; antennae and legs pale yellow-brown ; punctuation fine and (especially on elytra) very close. Thorax with fine, raised central line ; elytra with indistinct traces of striae, their raised basal margin finely toothed. L. ¾—1 l. Rare.   *P. pubescens*, Fab.

Phloeophthorus.

Very similar to *Carphoborus pilosus* but distinguished (apart from the generic characters) by the much larger and thicker club of antennae, the long, distinct, not scalelike pubescence on the longer and narrower thorax, the much broader rows of punctures on the elytra, with longer and stouter bristles on the much narrower interstices. L. 1 l. Rather common.   *P. rhododactylus*, Marsh.

Hylesinus.

A. Upper-side black, without hairs or only scantily covered with bristle hairs (at most standing more closely along suture).

a. Length 2—2½ lines.

Black or pitch-brown, the greatest part without hairs. Thorax narrowed in front, somewhat broader than long, closely and rather coarsely punctured; elytra with deep striae, interstices with sharp prominences placed in rows and with short blackish bristles. Common.

H. crenatus, Fab.

b. Length 1 line.

Black, dull; antennae and legs rust-brown. Thorax narrowed in front, much broader than long, punctured in wrinkles; elytra with punctured striae, interstices granulate, scantily covered with yellow, shiny bristles, placed more closely along suture. Not common.

H. oleiperda, Fab.

B. Upper-side closely covered with short scale- or bristle-like hairs.

a. Elytra without white spot from shoulder to middle of suture.

Black; elytra pitch-black or brown, with brown scales· spotted with gray-yellow hairs; antennae red-brown; legs (except tarsi) dark. Thorax much broader than long, finely granulate, without central line; elytra with distinct, fine rows of punctures, interstices with a row of small depressions. L. 1¼—1½ l. Common.   *H. fraxini*, Fab.

b. Elytra with white spot from shoulder to middle of suture.

Black; elytra pitch-black or brown, with brown and

yellow scales and also with an angular band of white scales reaching from shoulder of each elytron to middle of suture and usually enclosing a dark spot at junction; antennae red-brown; legs reddish-yellow-brown. L. ¾—1 l. Rather common. *H. vittatus*, Fab.

Cissophagus.

Lighter or darker brown; elytra lighter; legs yellow-brown; upper-side closely-covered with small, yellowish scale-like hairs. Thorax rather longer than broad, somewhat constricted in front, broadest in middle, with a slightly raised central line; elytra somewhat dilated behind middle, with deep punctured striae, the punctures large, quadrangular, interstices granulate, with small, stiff bristle hairs. L. 1 l. Not common. *C. hederae*, Schmidt.

Carphoborus.

Under-side and head black; elytra, antennae and legs yellow-brown; thorax rather darker. More elongate than *C. hederae;* thorax longer and narrower, scale-like pubescence forming a distinct dorsal ridge, broadest in middle, very little constricted in front; elytra less abruptly rounded behind, punctures of striae less clearly defined, bristles on interstices less stout and long; anterior tibiae more triangularly dilated, with only two or three teeth at apex. L. 1 l. Rare. *C. pilosus*, Ratz.

Hylurgus.

A. Second interstice on elytra without tubercles on apical declivity.

Usually black; elytra brown; antennae and tarsi rust-red; often entirely brown or yellow-brown. Head with strong scattered punctures, with a raised central line on front part of forehead; thorax moderately narrowed in front, with scattered punctures, central line smooth; elytra rather broader than thorax, double as long as together broad, with very fine punctured striae, interstices somewhat wrinkled and each with a row of little tubercles, except on second on apical declivity. L. 1¾—2 l. Common.

H. piniperda, Lin.

B. Second interstice on elytra with tubercles on apical declivity.

Similar to *H. piniperda* but usually rather smaller; elytra with more delicate punctured striae, interstices more closely punctured, row of tubercles on second one continued to apex; first tooth on outer-side of hinder tibiae placed in middle, second far from apex. Rare; Braemar.

B. minor, Hart.

Hylastes.

A. Mesosternum not produced; third tarsal joint broad heart-shaped.

   a. Central line of thorax either punctured or impunctate, but not raised.

      I. Thorax with scarcely a trace of impunctate central line, sides rounded.

Black, antennae and legs red-brown. Very similar to *H. ater* but more compressed; rostrum with a deep impression on each side at apex; thorax very closely punctured, with scarcely a feeble trace of impunctate central line; elytra broader and more strongly punctured. L. $1\frac{1}{2}$—$1\frac{3}{4}$ l. Not common.   *H. cunicularius*, Ratz.

      II. Thorax with a small, impunctate central line, sides straight.

Black, antennae and legs red-brown. Rostrum impressed at apex; thorax longer than broad, closely punctured, with impunctate central line on hinder half; elytra with punctured striae, interstices granulate and wrinkled. L. $1\frac{3}{4}$—2 l. Common.   *H. ater*, Payk.

   b. Thorax with a fine but distinct raised central line.

      I. Rostrum without longitudinal furrow.

Oblong. Black, dull; base of antennae and tarsi rust-red. Thorax almost transverse, strongly punctured; base of elytra not raised. L. $1\frac{1}{2}$ l. Rather common.

H. opacus, Er.

      II. Rostrum with a longitudinal furrow.

Black, dull; antennae and legs rust-red; with fine gray hairs, arranged in rows on elytra, rather more crowded at apex. Forehead finely and very closely punctured; rostrum usually with a shallow excavated central line; elytra with punctured striae, granules on interstices irregular in front and arranged in rows toward apex only. Narrower and more elongate than *H. opacus*, punctuation of thorax rather coarser, space between punctures shiny, tibiae a little more dilated.   L. 1½ l.   Rare.   *H. angustatus*, Herbst.

B. Mesosternum with a small prominence between intermediate coxae ; third tarsal joint dilated, bilobed.

a. Thorax reddish-brown.

Black ; thorax and elytra (except side margin) reddish-brown ; antennae and legs rust-red ; with fine gray hairs. Rostrum with a small, raised central line, sometimes with a feeble crescent-shaped furrow at base ; thorax strongly and very closely punctured, almost in wrinkles, with a very narrow, smooth central line ; elytra with deep punctured striae, interstices granulate and wrinkled.   L. 1¼—1½ l. Common.   *H. palliatus*, Gyll.

b. Thorax brown-black.

Short. Brown-black, obscure, slightly pubescent ; elytra dull red or pitchy ; legs and antennae dull red. Thorax closely and strongly punctured, with an obscure dorsal ridge ; elytra with deep punctured striae. L. 3¼ l. Rather common.   *H. obscurus*, Marsh.

CURCULIONIDAE.

A. Mentum leaving maxillae entirely uncovered.

   a. Anterior coxae placed more or less apart (*Gymnetron* excepted), prosternum generally with a furrow between them.

      I. Club of antennae solid ; tarsi imperfectly or not spongy beneath, third joint generally entire, with fourth joint inserted in depression on its upper surface.

         1. Pygidium covered by elytra.

Peduncle of submentum short, sometimes indistinct ; eyes at most moderately large, often very small or absent, standing far apart beneath, either finely or coarsely granulate ; club of antennae oval or nearly globular ; tibiae hooked at apex ; third tarsal joint often a little broader than second, rarely bilobed.   *Cossonides.*

         2. Pygidium not covered by elytra.

Peduncle of submentum reaching level of front of emargination ; eyes very large, standing almost always near each other beneath, finely granulate; club of antennae generally hatchet-shaped; tibiae hooked at apex; third tarsal joint usually much broader than second, never bilobed. *Calandrides.*

II. Club of antennae more or less distinctly articulate; tarsi usually spongy beneath, with third joint bilobed; peduncle of submentum not reaching level of front of emargination.

1. Epimera of mesothorax visible above between thorax and elytra; intermediate abdominal segments nearly always curved or angular at extremities.

A A. Scutellum distinctly visible.

Rostrum often thickened at base, scrobes turning quickly on to under surface and invisible on sides, except in front; prosternum somewhat rarely channeled; pygidium sometimes covered by elytra, sometimes uncovered; tibiae in most cases hooked at apex, but usually feebly; tarsal claws free or soldered. *Barididΐs.*

B B. Scutellum not or scarcely visible.

Rostrum not thickened at base, scrobes generally continued along lower part of side and visible wholly or in great part; prosternum very often channeled or hollowed out; pygidium not covered by elytra; tibiae not or scarcely at all hooked at apex; tarsal claws not soldered.

Ceuthorhynchides.

2. Epimera of mesothorax not visible above; intermediate abdominal segments straight behind.

A A. Antennae not elbowed.

Rostrum long, abruptly bent, forming an acute angle with head, received when at rest into a channel on prosternum, scrobes placed rather high, rounded, contiguous to eyes; channel on prosternum reaching mesosternum but not continued on to it; eyes very large, oval, contiguous on forehead; scutellum absent; elytra covering pygidium; posterior legs formed for leaping; tibiae not hooked at apex; tarsal claws simple. *Ramphides.*

B B. Antennae elbowed.

a a. Funiculus of antennae with six or seven joints.

Rostrum variable; prosternum with channel, rarely converted into a simple hollow; eyes almost invariably at least partly covered when rostrum is folded at rest; thorax generally with anterior margin prominent in middle or sinuate at sides, usually with ocular lobes; scutellum distinct or not; elytra nearly always covering pygidium; anterior coxae prominent; tibiae generally hooked at apex; tarsal claws simple.   *Cryptorhynchides.*

b b. Funiculus of antennae with five joints.

Rostrum slender, moderately long, cylindrical, sometimes gradually narrowed in front, scrobes oblique, quickly passing on to under-side; prosternum short, channeled or not, in latter case, anterior coxae contiguous; scutellum distinct; pygidium more or less uncovered; anterior coxae large, tolerably prominent; tibiae hooked at apex; tarsal claws variable.   *Gymnetrides.*

b. Anterior coxae placed close together, very rarely slightly separated, in which case prosternum not channeled.

I. Elytra leaving pygidium more or less uncovered, or if pygidium covered, tarsal claws appendiculate, split or toothed. (*Rhinomacer* excepted).

1. Intermediate abdominal segments angular at extremities.

A A. Funiculus of antennae with five joints.

Peduncle of submentum narrow and prominent; rostrum long, not very robust, cylindrical, scrobes commencing a little beyond middle, oblique; eyes placed moderately apart above; elytra imperfectly covering pygidium; tarsal claws simple, generally soldered.   *Cionides.*

B B. Funiculus of antennae with six or seven joints.

Peduncle of submentum tolerably prominent; rostrum long, slender, cylindrical, scrobes commencing more or less far from apex and reaching eyes, latter far apart above; elytra covering pygidium or not; tarsal claws appendiculate.   *Tychiides.*

2. Intermediate abdominal segments not angular at extremities.

Funiculus of antennae with six or seven joints.

A A. Antennae elbowed (in *Magdalinus* imperfectly).

   a a. Posterior angles of thorax at most right angles, not prominent; body not cylindrical.

      A a. Body not rhomboidal; prosternum very short before anterior coxae (in *Acalyptus* of ordinary length, but tarsal claws simple).

Peduncle of submentum slender and rather prominent; mandibles very short; rostrum not very robust, often slender, cylindrical, variable in length, scrobes commencing near or somewhat before middle, linear and reaching base; eyes nearly always rounded; elytra leaving pygidium slightly uncovered or not; tibiae not hooked; first tarsal joint generally short; claws variable.   *Anthonomides.*

      B b. Body rhomboidal; prosternum long before anterior coxae.

Peduncle of submentum slender and prominent; mandibles elongate triangular, prominent; rostrum very long and slender, scrobes commencing a little beyond middle, linear, reaching base; eyes transverse; elytra generally leaving pygidium slightly uncovered; tibiae not hooked (except sometimes anterior pair); first tarsal joint moderately long; claws bifid or appendiculate. *Balaninides.*

   b b. Posterior angles of thorax acute, more or less prominent; body cylindrical.

Peduncle of submentum tolerably long; mandibles very short; rostrum tolerably long, cylindrical, scrobes commencing in or slightly beyond middle, linear, reaching base; eyes transverse; elytra not covering pygidium; tibiae hooked at apex; tarsal claws simple.
          Magdalinides.

B B. Antennae not elbowed.

   a a. All abdominal segments free.

Peduncle of submentum variable in length; rostrum generally long, slender and dilated in front, scrobes shallow, commencing in middle, generally linear, reaching base; pygidium covered or not; intermediate coxae contiguous or separated by process of mesosternum; tibiae not hooked at apex; tarsal claws bifid or free.   *Rhinomacerides.*

b b. First two abdominal segments soldered together.

Peduncle of submentum broad and tolerably long; head elongate behind eyes, rostrum more or less robust and dilated in front, scrobes placed high, broad, deep; pygidium not covered, tibiae with one or two hooks at apex; tarsal claws soldered; fifth abdominal segment nearly always very small. *Attelabides.*

II. Pygidium covered by elytra.

1. Metasternum more or less long, its episterna generally at least moderately broad.

A A. Antennae not elbowed.

Peduncle of submentum slender and prominent; head more or less elongate behind eyes; rostrum long, curved, cylindrical, scrobes roundish, more or less distant from mouth; scape of antennae short, funiculus with seven joints, seventh free; eyes distant from thorax, nearly rounded; scutellum very small; apex of tibiae not hooked, truncate; third tarsal joint broader than second, claws free; first two abdominal segments soldered together; episterna of metathorax very narrow; epimera of mesothorax small. *Apionides.*

B B. Antennae elbowed.

a a. Tibiae compressed, bisinuate on inner-side, hooked (generally strongly) at apex.

Peduncle of submentum more or less prominent; rostrum almost always at most moderately robust, rounded at angles or cylindrical, scrobes linear, deep, straight, oblique; funiculus of antennae with seven joints, seventh often attached to club; eyes transverse; base of thorax generally bisinuate; tarsal claws free or soldered; insects nearly always winged. *Hylobiides.*

b b. Tibiae more or less rounded and not sinuate on inner-side.

A a. Tarsal claws nearly always soldered; antennae imperfectly elbowed.

Peduncle of submentum more or less prominent; rostrum variable, scrobes linear, deep, turning on to under surface and often meeting behind; funiculus of antennae either with joints gradually enlarging and merging in club or with seventh joint contiguous to club ; eyes transverse ; thorax nearly always with ocular lobes ; tibiae slightly hooked at apex ; tarsal claws nearly always soldered at base ; metasternum tolerably often short. *Cleonides.*

    B b. Tarsal claws free ; antennae completely elbowed.

    A 1. Rostrum more or less slender, cylindrical.

Peduncle of submentum prominent ; scrobes commencing generally far from mouth, linear, straight or oblique, reaching eyes ; funiculus of antennae with from five to eight joints, seventh nearly always free ; thorax very rarely with ocular lobes, scarcely ever bisinuate at base ; scutellum distinct ; tibiae usually slender and rounded, often hooked at apex ; tarsal claws very rarely soldered ; epimera of mesothorax small or moderately large ; insects nearly always winged. *Erirhinides.*

    B 1. Rostrum tolerably robust, rounded at angles.

Peduncle of submentum broad and more or less prominent ; rostrum longer than head, scrobes reaching mouth or nearly so, linear ; funiculus of antennae with seven, rarely six joints ; eyes nearly always transverse ; tibiae not hooked at apex ; tarsal claws free ; epimera of mesothorax usually moderately large ; insects generally winged.

    Hyperides.

    2. Metasternum very short, its episterna narrow.

Epimera of mesothorax much smaller than its episterna, acute in front ; tarsal claws free ; insects apterous.

    A A. Submentum with peduncle.

    a a. Tibiae hooked at apex.

Peduncle of submentum tolerably prominent ; rostrum more or less long, variable in stoutness, rarely angular, scrobes generally reaching mouth, or nearly so, linear, deep, reaching eyes ; thorax with very feeble (often rather indistinct) ocular lobes. *Molytides.*

    b b. Tibiae not or scarcely hooked at apex.

Peduncle of submentum short and broad, sometimes not very distinct ; rostrum moderately broad and long, angular or nearly so, very rarely rounded at angles and not parallel, scrobes complete in front, reaching eyes or nearly so ; thorax with or without ocular lobes, prosternum nearly always emarginate in front.   *Rhyparosomides.*

B B. Submentum without peduncle.

Rostrum more or less robust, ·received at rest into a channel on prosternum, scrobes linear, curved ; thorax with prominent ocular lobes, entirely covering eyes when at rest ; tibiae not hooked at apex ; tarsi narrow, not spongy beneath ; third joint not broader than second ; metasternum very short.   *Byrsopsides.*

B. Maxillae wholly or for the greatest part covered by mentum.

Submentum without peduncle or with only a trace of it ; rostrum more or less robust, scrobes reaching mouth or nearly so.

> *a.* Thorax with ocular lobes ; eyes generally large, transverse, pointed beneath and at least partly covered when at rest.

Scrobes linear, turning on to under-side ; antennae elbowed.   *Leptopsides.*

> *b.* Thorax without ocular lobes ; eyes round or short oval, not covered.

> *I.* Scrobes variable, but never both linear and directed downward.

Antennae elbowed, scape reaching beyond eyes behind, funiculus with seven (very rarely six) joints.
                    Otiorhynchides.

> *II.* Scrobes linear, directed downward.

Antennae elbowed, scape variable in length, funiculus with seven (rarely six) joints.   *Brachyderides.*

COSSONIDES.

A. Funiculus of antennae with seven joints.

> *a.* Scape of antennae reaching eyes or nearly so, but not reaching on to them ; second joint of funiculus very short.

> *I.* Eyes prominent.

Rostrum about as long as head, cylindrical, straight or nearly so, scrobes narrow, commencing beyond middle, oblique or curved; antennae usually inserted between base and middle of rostrum, club small; scutellum very small; anterior legs placed slightly apart; mesosternum narrow, linear. *Rhyncolus*, Creutz.

II. Eyes depressed.

Body more cylindrical than in *Rhyncolus*, obtusely rounded behind; rostrum short and nearly parallel-sided in male, shorter and nearly triangular in female; antennae inserted considerably behind middle of rostrum, shorter than in *Rhyncolus* (especially scape), glabrous, joints of funiculus closely articulated, club solid, compressed, cut off straight at apex; thorax convex, not constricted in front; femora (especially anterior pair) much thickened, with traces of a tooth beneath; front pairs of coxae nearly contiguous.

Stereocorynes, Woll.

b. Scape of antennae reaching on to eyes, sometimes extending beyond them; second joint of funiculus elongate.

I. Anterior coxae placed only slightly apart.

Similar to *Rhyncolus* but with antennae distinctly more slender, funiculus less closely jointed and less thickened, club larger. *Phloeophagus*, Schoenh.

II. Anterior coxae placed rather far apart.

1. Antennae inserted before middle of rostrum in male, near base in female.

Rostrum longer in male than in female, scrobes in former commencing in middle, in latter at base, very short. Slightly less cylindrical than *Mesites*, a little more convex, with eyes more approximated, antennae more elongate, club much larger and abrupt, legs slightly longer and somewhat farther apart, third tarsal joint minutely bilobed, thorax less oblong, body often diffusely covered with very delicate silky pubescence, and rostrum of male considerably longer and more slender, proportionally a little more widened at base of antennae. *Rhopalomesites*, Woll.

2. Antennae inserted before middle in both sexes.

Rostrum contracted behind, expanded in front; scrobes commencing far in front, abruptly curved; club of antennae large, more or less velvety; mesosternum tolerably broad; thorax more or less longitudinally impressed; third tarsal joint not dilated; eyes oval, not far apart above; body narrow, parallel-sided, bare.   *Cossonus*, Clairv.

B. Funiculus of antennae with five joints.

Rostrum a little longer than head, tolerably robust, cylindrical, slightly curved, scrobes commencing in middle, deep, reaching under margin of eyes; scape of antennae reaching on to eyes, second joint of funiculus sometimes elongate, club small; scutellum indistinct; anterior legs placed moderately apart; mesosternum tolerably broad.

Pentarthrum, Woll.

Rhyncolus.

A. Body more or less broad.

a. Interstices on elytra with a very fine, indistinct row of punctures.

Black, shiny; antennae and legs rarely pitch-black, usually brown. Rostrum scarcely so long as head, scrobes curved under eyes; thorax distinctly longer than broad, somewhat narrowed in front, with scattered punctures; elytra scarcely broader and almost double as long as thorax, with deep punctured striae. L. 1½ l. Rather common.

R. chloropus, Fab.

b. Interstices on elytra with fine, scattered punctures.

Dark brown; antennae and legs lighter. Rostrum rather longer and narrower than head, with a short central furrow; thorax as long as broad, rounded at sides, constricted in front; elytra with punctured striae. L. 1½ l. Common.

R. cylindrirostris, Ol.

B. Body elongate, linear.

Elongate, rather flat. Pitch-brown, shiny, bare; antennae and legs brown-red. Rostrum scarcely shorter than thorax, oblong, cylindrical, slightly curved, shiny, very finely punctured; vertex of head impunctate; thorax with large punctures, central line impunctate; elytra a little broader than thorax, strongly rounded at apex, feebly bordered, with deep punctured striae, interstices punctured almost in rows. L. 1¾ l. Rare. *R. gracilis,* Rosenh.

Stereocorynes.

Pitch-black or brown, shiny; antennae and legs lighter. Rostrum rather closely punctured; thorax as long as broad, sides slightly rounded, disc evenly, not closely but deeply punctured; elytra scarcely broader than, and scarcely double as long as thorax, with deep, coarse punctured striae, interstices smooth and shiny, with a not altogether regular, very fine row of punctures; anterior femora strong, somewhat compressed, dilated on each side. L. 1½ l. Rather common. *S. truncorum*, Germ.

Phloeophagus.

A. Body pitchy, with bronze reflection.

Broad convex. Pitchy, with a bronze reflection, shiny, smooth; club of antennae ovate; rostrum longer than head, robust, finely and closely punctured; thorax nearly oval, not very strongly and rather diffusely punctured; elytra scarcely dilated at sides, with a slight longitudinal impression on shoulders, with deep punctured striae, interstices indistinctly punctured. L. 1½ l. Moderately common.
P. aeneopiceus, Boh.

B. Body pitch-black.

Somewhat convex. Pitch-black, rather shiny, with fine gray pubescence; antennae and legs reddish. Rostrum not very robust, almost as long as thorax, closely punctured; thorax oblong, with large but not deep punctures; elytra slightly dilated at sides, with notched striae, interstices very finely, diffusely punctured and finely wrinkled transversely. L. 1¾ l. Not common. *P. spadix*, Herbst.

Rhopalomesites.

Elongate. Pitch-black, not very shiny, smooth; antennae and legs dark reddish. Rostrum moderately curved; thorax, oblong half as long again as broad, strongly constricted before apex, rather closely, moderately strongly punctured, slightly impressed before scutellum, narrowly bordered in front; elytra with moderately strong punctured striae, interstices flat, finely wrinkled. L. 3—4½ l. Not uncommon. *R. Tardii*, Curt.

Cossonus.

Black, bare; elytra, antennae and legs pitchy. Rostrum

more than double as long as head, rather thin, moderately dilated at apex; thorax nearly quadrangular, finely and not very closely punctured, not flat on disc, very indistinctly ridged, base scarcely impressed; elytra with deep, punctured striae, interstices impunctate.   L. 2½—3 l. Common.   *C. ferrugineus*, Clairv.

C. linearis, Lin. is smaller, with rostrum shorter, thicker and more dilated toward apex, thorax flat on disc, more deeply and coarsely punctured, with two longitudinal impressions at base, and tibiae less sinuated on inner-side.

Pentarthrum.

Narrow, nearly cylindrical.   Reddish-pitchy, rather shiny, bare; antennae and legs lighter. Rostrum punctured deeply at base, feebly toward apex; thorax elongate, strongly punctured, broadest near base; elytra with deep, roughly punctured striae, interstices with a row of very small punctures.   L. 1¾ l.   Rare.   *P. Huttoni*, Woll.

CALANDRIDES.

Rostrum variable in length, slightly curved, thickened at base, cylindrical in front, scrobes placed near base on lower part of sides, short; thorax distinctly longer than broad, a little shorter than elytra; club of antennae oval or oblong oval, episterna of metathorax moderately broad, its epimera small; epimera of mesothorax visible from above; body narrow, linear.   *Calandra*, Clairv.

Calandra.

A. Thorax with large, oblong punctures and impunctate central line.

Brown, rarely black; antennae and legs rust-red; elytra with deep punctured striae, interstices impunctate, alternate ones somewhat raised at base.   L. 1½—1¾ l.   Common.   *C. granaria*, Lin.

B. Thorax very closely punctured, punctures round and deep, impunctate central line indistinct.

Pitch-black, dull; a spot at shoulder, another behind middle of each elytron and side margin reddish. Elytra with extremely close punctured striae, interstices narrow,

alternate ones with very short yellowish bristles. L. $1\frac{1}{4}$ l.
Common. *C. oryzae*, Lin.

BARIDIDES.

Rostrum variable, but never very long or slender, often
separated from forehead by a transverse furrow, more or
less laterally compressed, curved; scrobes commencing in
middle or a little beyond it; antennae short, club large,
oblong oval or oval; prosternum not channeled; mesosternum not forming a continuous surface with pro- and
meta-sternum; pygidium not covered, small, nearly horizontal; body oblong. *Baris*, Germ.

Baris.

A. Upper-side not clothed with scales.

a. Under-side thickly covered with hairs.

Long, almost cylindrical. Black, moderately shiny;
upper-side with very fine, scattered, whitish hairs; meso-
and metasternum and abdomen thickly covered with white
scales. Thorax closely and deeply punctured, with smooth
central line; elytra with deep striae, which are scarcely
punctured, interstices with a row of very fine, feeble punctures. L. $1\frac{3}{4}$—2 l. Rather common. *B. T-album*, Lin.

b. Under-side without scales, or only with extremely
small bristles in the punctures.

I. Elytra unicolorous black.

Oblong. Black, rather shiny. Rostrum rather thick,
punctured; thorax oblong, very finely, shallowly, diffusely
punctured; elytra with very fine, shallow striae, in which
are small, isolated punctures, interstices flat, broad, with an
extremely fine network of scratches and a row of very fine
punctures. L. 2 l. Scarce. *B. laticollis*, Marsh.

II. Elytra unicolorous blue or green.

1. Thorax black; interstices on elytra distinctly
punctured.

Oblong. Black, not very shiny; elytra dark blue.
Rostrum as long as thorax, punctured; thorax not very
convex, finely and closely punctured, narrowed in front;
elytra with simple striae, not dilated behind shoulders,

interstices flat, distinctly punctured. L. 1½ l.   Scarce.
B. abrotani, Germ.

2. Thorax blue ; interstices on elytra feebly punctured.

A A. Thorax diffusely punctured, in middle
almost impunctate.

Oblong-ovate.   Black; thorax and elytra blue.   Thorax
somewhat longer than broad, narrowed in front; elytra
somewhat dilated behind shoulders, with fine, indistinctly
punctured striae, interstices with a row of feeble punctures.
L. 1⅓ l.   Rather common.   B. lepidii, Germ.

B B. Thorax closely punctured, with a narrow
central line almost impunctate.

Oblong-ovate.   Black ; thorax and elytra blue-green.
Thorax narrowed in front; elytra short, with prominent
shoulders, with fine, deep striae, outer ones indistinctly
punctured, interstices with a row of very fine indistinct
punctures.   L. 1⅓ l.   Devizes.   B. chlorizans, Germ.

III. Elytra black at base, red at apex.

Oblong.   Black ; apex of elytra red.   Rostrum curved ;
thorax tolerably closely punctured, indistinctly bordered in
front; elytra with notched striae, interstices indistinctly
punctured.   L. 1¾ l.   Rare.   B. analis, Ol.

B. Upper-side covered with scales.

Black or black-brown ; not very shiny ; sides of thorax,
the elytra and under-side checkered with white and brown-
yellow scales.   Rostrum as long as head and thorax, punctured and furrowed ; thorax very closely punctured, without
smooth central line ; elytra with fine striae.   L. 1¼—1½ l.
Rare.   B. scolopaceus, Germ.

CEUTHORHYNCHIDES.

A. Eyes completely uncovered.

a. Funiculus of antennae with six joints ; rostrum
elongate, moderately robust.

Scrobes commencing a little before middle of rostrum,
very oblique ; first three joints of funiculus of antennae
elongate ; prosternum not hollowed out before anterior
coxae, entire in front ; anterior coxae placed slightly but

distinctly apart; tarsal claws toothed at base; pygidium not covered. *Amalus*, Schoenh.

b. Funiculus of antennae with seven joints; rostrum at most moderately long, rather robust.

I. Prosternum strongly emarginate in front.

Scrobes commencing far forward, narrow and oblique; first two joints of funiculus of antennae elongate; eyes generally with a very short orbit above; prosternum broadly and rather strongly hollowed out before anterior coxae; latter standing farther apart than in *Phytobius*; tarsal claws toothed at base; pygidium not covered.

Rhinoncus, Schoenh.

II. Prosternum feebly emarginate in front.

Scrobes commencing near apex of rostrum, very oblique; first three joints of funiculus of antennae elongate; eyes often with a very short orbit above; prosternum more or less hollowed out before anterior coxae; latter placed more or less apart; tarsal claws simple or toothed at base; pygidium not covered. *Phytobius*, Schoenh.

B. Eyes at least partly covered by thorax.

a. Prosternal channel for reception of rostrum effaced between anterior coxae, sometimes absent.

I. Body oblong or oblong oval; metasternum moderately long.

1. Funiculus of antennae with six joints; tarsal claws toothed at base.

Rostrum rather long and robust; scrobes commencing a little before middle of rostrum; eyes nearly round; first three joints of funiculus of antennae elongate; pygidium not covered. *Tapinotus*, Schoenh.

2. Funiculus of antennae with seven joints; tarsal claws simple.

Rostrum elongate, slender, scrobes commencing toward middle of rostrum; eyes short oval; first two joints of funiculus of antennae elongate; pygidium not covered.

Poophagus, Schoenh.

II. Body thick, very short; metasternum very short. Tarsal claws appendiculate or bifid, very rarely simple.

1. Funiculus of antennae with six joints.

A A. Thorax with ocular lobes.

Rostrum elongate, moderately robust ; scrobes commencing about one-third of length from apex of rostrum ; first three joints of funiculus of antennae elongate ; thorax nearly as long as broad ; elytra very convex ; pygidium uncovered.   *Rhytidosomus*, Schoenh.

B B. Thorax without ocular lobes.

Similar to *Ceuthorhynchus* except in funiculus of antennae.   *Ceuthorhynchideus*, Duv.

2. Funiculus of antennae with seven joints.

Rostrum generally slender, scrobes commencing in or near middle ; first two joints of funiculus of antennae elongate, rest gradually decreasing in length ; thorax rarely as long as broad ; elytra moderately or only slightly convex ; pygidium uncovered.   *Ceuthorhynchus*, Germ.

b. Prosternal channel for reception of rostrum continued on to mesosternum.

I. Fourth tarsal joint moderately long, with two claws.

Rostrum elongate, at most moderately robust ; scrobes commencing near middle ; funiculus of antennae with seven joints, first two elongate, nearly equal, third and fourth longer than last three ; tarsal claws more or less toothed at base ; pygidium not covered.   *Coeliodes*, Schoenh.

II. Fourth tarsal joint very short, with only one claw.

Head with fine central ridge, forehead slightly impressed ; rostrum elongate, not very robust, scrobes commencing near middle ; funiculus of antennae with seven joints, first two elongate, first longer than second, the other five short, nearly equal ; pygidium not covered.   *Mononychus*, Germ.

Amalus.

Short ovate, convex. Black ; under-side with white scales ; legs and base of antennae rust-red ; thorax and elytra strewn with very small grayish scales, latter red-brown at apex or entirely brown, the grayish scales rather closer at suture and forming a spot at base. Thorax strongly and closely punctured; elytra with deep punctured striae, interstices narrow, even.   L. ¾ l.   Not uncommon.

A. scortillum, Herbst.

Rhinoncus.

A. Thorax with two (or without any) tubercles.

a. Central furrow on thorax indistinct.

I. Interstices on elytra wrinkled, or granulated in wrinkles.

1. Body short ovate; elytra not, or only slightly longer than together broad; femora clubbed toward apex.

Black; antennae and legs brown or rust-red; under-side and lateral margin of elytra thickly covered with white scales; upper-side with fine gray hairs, base of suture with a white spot. Thorax very closely and strongly punctured, with a small, indistinct tubercle on each side; elytra with distinct punctured striae, interstices finely granulated in wrinkles. L. 1¼—1½ l. Common. *R. pericarpius*, Fab.

2. Body ovate; elytra distinctly longer than together broad; femora slightly thickened near middle.

Black, sprinkled with white; under-side and an oblong spot at scutellum covered with white scales; base of antennae, tibiae and tarsi (sometimes also femora) yellow-brown. Thorax almost cylindrical, central line and a line on each side more thickly covered with white scales; elytra with deep punctured striae, interstices narrow, somewhat wrinkled. L. 1—1¼ l. Rather common.

R. subfasciatus, Gyll.

II. Interstices on elytra with distinct tubercles.

Black; antennae and legs rust-red; under-side thickly covered with white scales, sides of thorax and its central line with whitish ones, elytra speckled with small grayish spots (especially behind), with an oblong white, or whitish-yellow, oblong spot at base of suture. Thorax with a blunt tubercle on each side in middle, strongly and closely punctured; elytra with striae, the side ones distinctly punctured, interstices rough, with many small tubercles behind. L. 1 l. Common. *R. castor*, Fab.

b. Central furrow on thorax distinct.

I. Interstices on elytra finely wrinkled.

Brown-black ; antennae and legs yellow-brown ; upperside checkered with fine gray and reddish scalelike hairs, base of suture lighter, under-side with white scales. Thorax with a small pointed tubercle on each side. L. 1¼ l. Rather common.   *R. inconspectus*, Herbst.

II. Interstices on elytra with a few scattered tubercles.

Short ovate, not very convex. Black; antennae and legs reddish-yellow; under-side thickly covered with white scales ; thorax sometimes with three white lines and elytra speckled with white. Thorax with two tubercles, apical margin scarcely raised ; elytra with more distinct punctured striae than in *R. castor*. L. ¾—1 l. Moderately common.   *R. bruchoides*, Herbst.

B. Thorax with four tubercles.

Oblong-ovate. Black; tibiae red-yellow ; under-side and base of suture of elytra thickly covered with grayish-white scales ; thorax and elytra sprinkled with white scales. Apical margin of thorax scarcely raised, but with two small tubercles in middle, surface closely granulate, with an indistinct central furrow and with blunt tubercles on each side behind ; elytra with moderately deep punctured striae, interstices with a fine network of scratches. L. ¾—1 l. Rare.   *R. denticollis*, Gyll.

Phytobius.

A. Antennae inserted in middle of rostrum ; fourth tarsal joint as long as all the rest together.

a. Tibiae and tarsi scantily fringed ; thoracic tubercles strong.

Black, dull ; antennae and legs (except knees) reddish-yellow ; under-side, head, sides and central line of thorax, and sides and suture of elytra closely, rest of upper-side more scantily covered with yellowish or greenish-gray scales. Thorax much broader than long, narrowed in front, with two very small, pointed tubercles at anterior margin and a much larger tubercle on each side near base ; elytra much broader than thorax, with fine but deep punctured striae, interstices broad, the fifth more raised than

the others, especially at base. L. 1 l. Rather common.
P. *velatus*, Beck.

b. Tibiae and tarsi not fringed ; thoracic tubercles not strong.

Black ; base of antennae and the legs (except knees and tarsi) reddish-yellow ; under-side, sides of thorax and a spot at base of suture of elytra closely covered with white scales ; upper-side black-brown, with a silvery reflection. Thorax with two extremely indistinct tubercles at anterior margin and a large, pointed tubercle on each side at base ; elytra as in *P. velatus.* L. 1¼ l. Rather common.
P. *leucogaster*, Marsh.

B. Antennae inserted before middle of rostrum ; fourth tarsal joint not very elongate.

a. Club of antennae oval, pointed, scrobes straight ; tarsal claws simple.

I. Thoracic tubercles distinct.

1. Interstices on elytra not roughened behind.

A A. Thorax with distinct central furrow.

Short ovate. Black ; tibiae rust-brown ; thorax with a broad line of white scales on each side ; elytra with spots of whitish scales, base of suture impressed, velvety-black. Thorax with four tubercles and a central furrow ; elytra with punctured striae, interstices convex. L. 1 l. Not common. P. *notula*, Germ.

B B. Thorax without central furrow.

Short. Black, rather convex; antennae and legs yellow; femora with a black-brown spot before apex beneath; a broad streak on each side of thorax and the under-side closely covered with white scales; elytra variegated with white. Thorax closely and moderately strongly punctured, with four not very strong tubercles, without central furrow; elytra with strong punctured striae, interstices not rough. L. 1 l. Not uncommon. P. *Waltoni*, Boh.

2. Interstices on elytra roughened behind.

Short ovate. Black ; upper-side spotted, under-side and sides closely covered with white scales ; antennae red-brown, tibiae and tarsi yellow. Forehead depressed ; thorax uneven, with four distinct tubercles, disc rather bare ; elytra with punctured striae, interstices convex, shagreened, rough behind.   L. 1 l.   Rather common.

P. quadrituberculatus, Fab.

II. Thoracic tubercles indistinct.

Short ovate.   Black ; sides of thorax and the under-side closely covered with whitish scales ; scape of antennae, tibiae and tarsi reddish-yellow ; elytra with scattered spots of whitish scales toward apex.   Thorax with a feeble central furrow and an indistinct tubercle on each side ; elytra with deep, rather indistinctly punctured striae. L. 1 l.   Not uncommon.   *P. comari*, Herbst.

b. Club of antennae oblong oval, blunt, scrobes strongly curved ; each tarsal claw split.

I. Thorax with central furrow.

Short oval, convex.   Black; antennae and legs red ; base of suture and under-side closely covered with whitish scales.   Thorax deeply punctured, anterior margin excised; elytra with interstices of striae roughened at sides. Similar to *P. quadricornis*, but not convex, with thorax more coarsely and deeply punctured.   L. ¾ l.   Rare.

P. quadrinodosus, Gyll.

II. Thorax without central furrow.

Black ; upper-side scantily, a spot at base of suture of elytra and the under-side closely covered with whitish scales ; antennae and legs red-brown.   Thorax deeply and strongly punctured, with apex deeply emarginate and with a pointed tubercle on each side before base, apical tubercles remote ; elytra with deep, rather indistinctly punctured striae, interstices somewhat roughened at sides.   L. ⅘ l. Scarce.   *P. quadricornis*, Gyll.

Tapinotus.

Black or pitch-brown ; closely covered with whitish scales ; antennae and legs usually lighter ; thorax with two broad brown streaks, separated by a white central line ; elytra with a common black transverse spot on middle, with striae and a callosity before apex.   L. 2 l.   Very rare.   *T. sellatus*, Fab.

Poophagus.

A. Femora not toothed.

Oblong, rather flat. Black ; closely covered with whitish scales ; disc of thorax brownish, with white, central line ; elytra with five, not sharply defined dark spots, viz., a common one in middle and two others on each. L. 1—1½ l. Rather common. *P. sisymbrii*, Fab.

B. Femora with a little tooth.

Oblong, rather flat. Greenish-bronze ; under-side closely, upper-side more scantily covered with gray scales ; apex of rostrum, antennae and tarsi red-brown ;. elytra with green reflection. L. 1—1½ l. Scarce. *P. nasturtii*, Germ.

Rhytidosomus.

Black ; sides of breast and usually also a part of suture of elytra covered with white scales. Thorax strongly and closely punctured, anterior margin somewhat raised ; elytra with deep, punctured furrows, interstices narrow, with small granules. L. ¾ l. Scarce. *R. globulus*, Herbst.

Ceuthorhynchideus.

A. Femora not toothed.

  a. Elytra unicolorous.

    I. Upper-side more or less closely covered with whitish-gray scales.

      1. Base of thorax bisinuate.

        A A. Thorax with a small but distinct tubercle on each side.

          a a. Legs black.

Black ; closely covered on upper-side with gray scales (lighter at suture) and on under-side with whitish ones. Thorax much constricted before apex, with feeble central furrow (deeper in front and behind), anterior margin raised ; elytra with deep punctured striae, interstices narrow, somewhat convex. L. ¾ l. Common. *C. floralis*, Payk.

          b b. Femora brownish, tibiae and tarsi red-yellow.

Ovate, rather flat. Brown, dull ; covered above and beneath with grayish scales ; elytra pale brown ; antennae and rostrum black. Thorax constricted in front, with indistinct central furrow, anterior margin raised ; elytra with narrow striae, interstices flat, with a fine network of scratches. L. ¾ l. Rare.   *C. hepaticus*, Gyll.

B B. Thorax with an extremely indistinct tubercle on each side.

Short ovate. Black, under-side rather closely, head and thorax more finely and scantily covered with whitish scales, interstices on elytra with two more or less regular rows of grayish scales, base of suture closely covered with white scales. Thorax much constricted before apex, central furrow base only, anterior margin much raised; elytra with distinct at obtusely prominent shoulders, and deep, broad striae. L. ⅔—¾ l. Rare.   *C. pulvinatus*, Gyll.

2. Base of thorax straight.

Similar to *C. floralis*, but with thorax shorter, flatter, and more narrowed in front, central furrow visible in front only; scales seldom lighter on suture of elytra. L. ¾ l. Not uncommon.   *C. nigrinus*, Marsh.

II. Upper-side scantily covered with grayish scales.

1. Rostrum red at apex, black at base; striae on elytra indistinctly punctured.

Black or black-brown, dull; upper-side scantily, under-side closely covered with white scales; rostrum red, black at base; tibiae reddish-yellow. Thorax finely punctured, much constricted before apex, without central furrow, with an indistinct tubercle on each side, anterior margin moderately raised; elytra with deep, indistinctly punctured striae, interstices convex. L. 1 l. Common.

C. pyrrorhynchus, Marsh.

2. Rostrum black or brown; striae on elytra distinctly punctured.

A A. Legs black.

Rather flat. Black; upper-side scantily covered with gray scales, under-side closely with whitish ones. Rostrum as long as head and thorax, thin, curved, punctured; thorax rather convex, closely punctured, constricted before apex, with an indistinct tubercle on each side; elytra with moderately strong punctured striae, interstices nearly flat, shagreened, shoulders raised and rounded. L. ¾ l. Not uncommon.   *C. melanarius*, Steph.

B B. Legs red-yellow.

Short ovate. Pitch-brown, scantily covered with white scales and short bristles; rostrum reddish-brown; legs red-yellow. Rostrum as long as half the body, thin, curved, very finely striate, shiny; thorax very short, deeply constricted before apex, with an indistinct tubercle on each side, anterior margin much raised; elytra much broader than thorax, scarcely longer than broad, very convex, with deep punctured striae, interstices narrow, somewhat wrinkled, shoulders raised, obtusely angular. L. ½ l. Rare.

C. pumilio, Gyll.

b. Elytra black, with apex reddish.

Black, with gray pubescence; apex of elytra, tibiae and tarsi yellow-red; base of suture of elytra with a spot of white hairs. Rostrum long, thin, curved, very finely striate, shiny; thorax short, deeply constricted toward apex, with indistinct tubercles on each side, middle of anterior margin raised; elytra with rather deep crenate striae; interstices transversely wrinkled. L. 1¼ l. Not uncommon.

C. terminatus, Herbst.

B. Femora toothed.

a. Anterior margin of thorax notched at sides; interstices of elytra roughened throughout.

Pitch-black; elytra, antennae and legs red-brown. Elytra with feebly punctured striae, interstices raised and bearing long white and black upright bristles and spines. L. 1¾ l. Moderately common. *C. horridus*, Fab.

b. Anterior margin of thorax simple; elytra roughened at apex only.

I. Thorax with three longitudinal gray lines; elytra without white spot at scutellum.

1. Rostrum unicolorous.

A A. Forehead without white spot.

Lighter or darker brown; covered with grayish scales beneath; antennae and legs reddish-yellow-brown; thorax without lateral tubercles and with three lines of white scales; elytra usually with blackish suture, upright whitish bristles on the interstices and some small spines before apex. L. 1 l. Common. *C. troglodytes*, Fab.

C. Chevrolati, Bris. (MS.) is a variety of *C. troglodytes* distinctly striped and banded with white.

_B B._ Forehead with white spot.

Oval, moderately convex. Brown, more or less reddish, rather dull, covered above scantily with grayish, beneath closely with whitish scales ; forehead with a spot of white scales ; thorax nearly conical, slightly constricted in front, with three gray longitudinal lines ; elytra yellow-red, with suture blackish and a transverse denuded band brownish, interstices convex, with rows of small recumbent hairs ; legs red-yellow. Antennae shorter than in _C. troglodytes_ ; punctuation close and not very fine.   L. $\frac{3}{4}$ l.   Rare.

C. frontalis, Bris.

_2._ Rostrum brown-red, with apex black.

Nearly oval, moderately convex. Brown-red ; antennae and legs red-yellow ; apex of rostrum and suture of elytra blackish ; covered above rather closely with gray, beneath closely with whitish scales. Thorax nearly conical, scarcely constricted in front, anterior margin scarcely raised, three gray longitudinal lines indistinct ; elytra with striae rather fine, interstices slightly convex, rather strongly roughened behind, with rows of very short recumbent hairs. L. $\frac{3}{5}$—$\frac{3}{4}$ l. Not common.   _C. Dawsoni_, Bris.

_II._ Thorax without gray lines ; elytra with white spot at scutellum.

_1._ Elytra short, with apical margin closely covered with white scales.

Short. Black, scantily covered with white scales ; thorax with lateral tubercles, anterior margin raised; elytra obtriangular-quadrate, much abbreviated, with white scales placed more closely in an obscure scutellary spot and along apical margin; tarsi yellow-red, last joint black at apex; antennae blackish ; punctuation nearly in wrinkles.   L. $1\frac{1}{4}$—$1\frac{1}{2}$ l. Rare.   _C. nigroterminatus_, Woll.

_2._ Elytra not short, apical margin not closely covered with white scales.

_A A._ Tarsi black.

Black, not very shiny, rather scantily covered with small dark scales, generally with a slight violet reflection, and strewn also with somewhat larger gray scales ; elytra with a spot of white scales at base of suture. L. 1 l.   Not common.   _C. versicolor_, Bris.

_B B._ Tarsi red-yellow.

Short oval, not very convex. Black, somewhat dull; tarsi red-yellow; covered rather scantily with brownish-yellow scales, with slight golden-yellow reflection, especially toward base of thorax, with whitish scales forming a spot at scutellum and scattered here and there on rest of thorax and elytra. Thorax with feeble lateral tubercles, very slightly constricted toward apex, anterior margin only slightly raised; elytra rather strongly narrowed behind. Thorax more flattened than in *C. versicolor*, with anterior margin less raised, tarsal claws smaller. L. $\frac{4}{5}$ l. Rare.

C. Crotchi, Bris.

Ceuthorhynchus.

A. Femora not toothed.

a. Suture of elytra white throughout.

Black, dull; base of antennae and the legs rust-red; under-side very closely covered with white scales; upper-side with gray and brown hairs, central line of thorax, suture and side margin of elytra closely covered with white scales. Anterior margin of thorax moderately raised, sides without tubercle. L. 1 l. Rare. *C. suturalis*, Fab.

b. Suture of elytra at most white at base.

I. Interstices of elytra with small tubercles before apex.

1. Thorax finely punctured.

A A. Elytra black.

a a. Body deep black, closely pubescent.

Similar to *C. assimilis* but deep black, without metallic reflection, upper-side much more thickly covered with hairlike scales; lateral tubercles on thorax feebler; elytra not much longer than together broad, tubercles before apex distinct. L. $1\frac{1}{4}$ l. Rare. *C. syrites*, Germ.

b b. Body leaden-black, scantily pubescent.

Leaden-black; under-side rather closely, upper-side scantily covered with hairlike scales. Rostrum long and thin; thorax much narrowed and constricted toward apex, anterior margin raised, with a central furrow (deeper before and behind, and more closely covered with scales) and with a small raised transverse line on each side, finely and very closely punctured; elytra one-third longer than

together broad, tubercles before apex feeble. L. 1—1¼ l.
Common. C. *assimilis*, Payk.
B B. Elytra blue.

Black, with some bronze reflection; elytra blue. Thorax
with anterior margin strongly raised and with a central
furrow (deeper in front and behind), and a small tubercle
on each side; elytra much broader than thorax, with strong
punctured striae; interstices flat, very finely wrinkled, with
rows of very fine, whitish recumbent hairs, tubercles before
apex pointed. L. ¾—1 l. Common. C. *erysimi*, Fab.

2. Thorax strongly punctured.

Black, shiny; elytra often with bluish or greenish
reflection; under-side very closely, upper-side very scantily
covered with grayish scales. Thorax narrowed and constricted before apex, with a more or less distinct central
furrow and a feeble tubercle on each side; elytra with deep
punctured striae, interstices wrinkled, with a row of fine
grayish hairlike scales, tubercles before apex pointed.
L. ½—¾ l. Common. C. *contractus*, Marsh.

II. Interstices of elytra without tubercles before
apex.

1. Legs black.

A A. Antennae black; thorax with a tubercle
on each side.

a a. Elytra without white spot at base of
suture.

Black; upper-side scantily covered with white bristles,
under-side with whitish scales. Rostrum scarcely as long
as head and thorax, curved; thorax narrower than in *C.
contractus*, more deeply and widely constricted before apex,
closely punctured, with a very feeble tubercle on each side;
elytra with deep punctured striae, interstices only slightly
convex, almost smooth. L. ¾ l. Scarce. C. *setosus*, Boh.

b b. Elytra with a spot of white scales at base
of suture.

Black; upper-side with scanty gray pubescence; base of
suture of elytra and the breast thickly covered with white
scales. Thorax deeply but not closely punctured, with an
indistinct central furrow and a blunt tubercle on each side,
base bisinuate; elytra rounded at shoulders, with deep

punctured striae, interstices somewhat wrinkled. L. ¾ l. Not uncommon. *C. cochleariae*, Gyll.

_B B._ Antennae brown; thorax without lateral tubercles.

Short ovate. Black, diffusely covered with white scales; antennae brown. Vertex of head ridged; thorax closely and distinctly punctured, with an indistinct central furrow, rather deeply constricted toward apex, anterior margin raised, base slightly bisinuate; elytra with deep punctured striae, interstices narrow, finely wrinkled. L. ¾ l. Scarce. *C. constrictus*, Marsh.

_2._ Legs red.

Black; sometimes powdered with yellowish-green; antennae, legs and rostrum reddish; base of suture of elytra and the breast covered with white scales. Rostrum very long and thin; thorax with scarcely any central furrow, with a tubercle on each side; elytra much broader than thorax, shoulders prominent, with deep punctured striae, interstices wrinkled with tubercles and with rows of gray hairs. L. 1 l. Common. *C. ericae*, Gyll.

_B._ Femora toothed.

_a._ Anterior margin of thorax distinctly notched or toothed at sides.

Black; under-side closely covered with white scales, upper-side with brown ones; thorax with posterior margin and three narrow lines, elytra with various confused lines white. Elytra with spines toward sides and at apex. L. 2 —2½ l. Moderately common. *C. echii*, Fab.

_b._ Anterior margin of thorax simple.

_I._ Outer margin of tibiae (somewhat before middle) with a large triangular tooth, thence cut obliquely to apex, with a row of bristles.

Black, dull; elytra with a spot at side and another, crescent-shaped, at apex formed of white scales; under-side thickly covered with white scales; legs red-brown. Thorax with a depression before scutellum, without lateral tubercles, central furrow very indistinct; elytra with feeble striae. L. 1¼ l. Scarce. *C. viduatus*, Gyll.

_II._ Tibiae somewhat dilated and rounded at apex, usually bearing bristles, but not toothed.

K 2

1. Ground colour of elytra black.

A A. Central furrow on thorax not deep.

    a a. Upper-side with gray or white scales, placed more thickly in some places than in others.

    A a. Antennae wholly or partly red-yellow or yellow-brown.

      A 1. Thorax strongly constricted before apex, anterior margin much raised.

       a 1. Lateral tubercles on thorax entirely surrounded by whitish scales.

Short ovate, rather flat. Black, dull ; antennae and tarsi yellow-brown ; sides of thorax (until beyond lateral tubercles), on elytra a common cruciform basal spot, an apical one and an abbreviated lateral band, and the underside of body covered with white scales. Thorax with central furrow, lateral tubercles bluntly conical ; elytra with punctured striae. L. 1¼—1½ l. Rather common.

                       C. litura, Fab.

       b 1. Lateral tubercles on thorax (if any) not surrounded by whitish scales.

      A 2. Suture of elytra without black spot.

       a 2. Elytra with distinct, well-defined white spots.

        A 3. Lateral tubercles on thorax distinct.

         a 3. Tibiae black.

Black, dull ; antennae and tarsi reddish ; sides of thorax, on elytra a broad apical spot, an abbreviated lateral band and a small composite spot near base of suture on each, and the under-side of body covered with white scales, base of suture with pale yellowish scales. Thorax flat, with central furrow ; elytra with punctured striae. L. 1½ l. Rather common.   *C. trimaculatus,* Fab.

         b 3. Tibiae yellow-brown.

Oblong-ovate, slightly convex. Black, dull ; under-side closely covered with white scales ; antennae, tibiae and tarsi dark yellow-brown ; sides and central line of thorax whitish ; elytra checkered with whitish, base of suture and (attached to this) a transverse band, bent forward, behind middle closely covered with white scales. Thorax somewhat shorter than broad, with a tubercle on each side, distinctly constricted in front; elytra twice as long as thorax, with moderately strong punctured striae. L. 1¼ l. Rare.

C. triangulum, Boh.

B 3. Lateral tubercles on thorax extremely indistinct.

Black, dull ; antennae, tibiae and tarsi red-brown ; elytra dark-brown, their apex, a spot near base of suture and an abbreviated lateral band (composed of three lines) and the under-side of body closely, rest of upper-side scantily covered with white scales. Thorax short, deeply but narrowly constricted before apex, with slight central furrow behind ; elytra with narrow, indistinctly punctured striae, interstices flat. L. 1—1¼ l. Rather common.

C. asperifoliarum, Gyll.

b 2. Elytra without distinct, well-defined spots.

Black, dull ; base of antennae and tarsi yellow-red ; upper-side scantily and unequally, under-side closely covered with white scales. Thorax closely and distinctly punctured, with fine central furrow and feeble lateral tubercles ; elytra with deep punctured striae, interstices narrow, somewhat convex. L. 1—1¼ l. Rare.

C. urticae, Boh.

B 2. Suture of elytra with a black spot, enclosed by white.

a 2. Tibiae red-brown ; length 1¼ —1½ lines.

Short ovate, rather flat. Black, dull ; antennae and tibiae reddish-brown ; tarsi red-yellow; thorax with sides and three lines on disc (outer ones slightly curved), elytra with numerous short lines (two on each near suture longer), suture with a common spot (interrupted in middle), and under-side of body closely covered with whitish scales. Lateral tubercles on thorax obtuse ; elytra with feeble punctured striae. Rather common. *C. campestris,* Gyll.

b 2. Tibiae yellow-brown ; length 1 line.

Short ovate, somewhat convex. Brownish-black, dull antennae, tibiae and tarsi yellow-brown ; disc of thorax with three parallel lines of grayish scales, various spots on elytra (sutural one interrupted), and under-side of body covered with whitish scales. Thorax short, with lateral tubercles. Not uncommon. *C. rugulosus*, Herbst.

B 1. Thorax slightly constricted before apex, anterior margin only slightly raised.

Oblong-ovate, somewhat convex. Black, dull ; antennae, tibiae and tarsi yellow-brown ; under-side, central line and sides of thorax, on elytra front half of suture, an oblique spot reaching from shoulder to middle of disc and the apex covered with white scales, interstices on elytra checkered with brown and gray. Thorax much broader than long, its lateral tubercles distinct ; elytra with fine striae, interstices even, with small black tubercles before apex. L. 1¼ l. Rare. *C. arcuatus*, Herbst.

B b. Antennae black.

A 1. Elytra with distinct markings.

a 1. Tibiae yellow-brown.

Oblong-ovate, somewhat convex. Black, dull ; antennae, tibiae and tarsi yellow-brown ; under-side closely covered with white scales ; thorax with sides and a central line, elytra with an oblique spot at shoulder, the apex and the suture (except middle) covered with white scales. Thorax scarcely broader than long, slightly constricted toward apex, anterior margin only slightly raised, with central furrow and lateral tubercles ; elytra with punctured striae. L. 1¼ l. Rather common. *C. melanostictus*, Marsh.

b 1. Tibiae black.

A 2. Tarsi black.

Black, slightly shiny ; thorax with lateral tubercles, and with a whitish central line ; elytra with strong punctured striae, strewn with white scales, condensed into a spot below scutellum, on sides toward middle and at apex. L. 1½ l. Scarce. *C. euphorbiae*, Bris.

B 2. Tarsi red.

Black ; tarsi red ; under-side rather closely, upper-side rather scantily covered with whitish scales, forming spots on elytra, base of suture closely covered with white or yellow scales. Thorax much narrowed in front, anterior margin strongly raised, with central furrow and distinct lateral tubercles; elytra with fine, scarcely punctured striae, interstices broad and flat, with raised granules behind. L. 1¼ l. Common. *C. quadridens*, Panz.

B 1. Elytra indistinctly variegated.

Rather flat. Black; tibiae and tarsi reddish; under-side closely, upper-side scantily covered with yellowish scales, placed more thickly at sides of thorax and in places on elytra, base of suture with white scales. Thorax with distinct central furrow, transversely somewhat convex in middle; elytra with narrow punctured striae, distinctly spined at sides and behind. L. 1¼ l. Not common.

C. resedae, Marsh.

b b. Upper-side evenly covered with brown, or brown and gray scales or hairs.

A a. Antennae black; base of suture of elytra with a white spot of scales.

A 1. Tibiae black.

a 1. Body scantily covered with brown and gray scales.

A 2. Thorax without lateral tubercle; length 1⅓—1½ lines.

Black, dull; under-side, a spot at base of suture of elytra and their side margin closely covered with white scales, rest of upper-side scantily covered with brown and gray ones. Thorax much broader than long, with slight central furrow, deeply constricted before apex, anterior margin much raised; elytra with narrow striae; pygidium rounded, entire. Moderately common. *C. marginatus*, Payk.

B 2. Thorax with a very small lateral tubercle; length 1 line.

a 2. Pygidium entire.

Similar to *C. punctiger*, but of shorter form, greater convexity and with relatively wider striae. London.

C. rotundatus, Bris.

b 2. Pygidium excised at apex.

Short ovate. Black; under-side, a spot at base of suture

of elytra and their side margin closely covered with scales. Similar to *C. marginatus*, but with fewer scales on upper-side, second joint of antennae much longer, thorax more widely constricted before apex and more convex, elytra narrower toward apex, sides not roughened, second tarsal joint transverse, rostrum of female longer and narrower. Not uncommon. *C. punctiger*, Gyll.

b 1. Upper-side tolerably closely covered with brown hairs.

Short ovate, somewhat convex. Deep black, shiny; tarsi pitchy; under-side, a short line in middle of front of thorax and common spot on elytra near scutellum closely covered with yellowish scales. Thorax with central furrow and lateral tubercles, deeply constricted toward apex, anterior margin rather strongly raised; elytra with tolerably deep punctured striae, interstices convex, wrinkled, spined toward apex. L. 1 l. Rare. *C. pilosellus*, Gyll.

B 1. Tibiae red-yellow.

Rather flat. Black, dull; tibiae and tarsi red-yellow; under-side and base and apex of suture of elytra closely covered with grayish scales, rest of upper-side scantily covered with brown scales. Thorax with indistinct central furrow, constricted toward apex, lateral tubercles conical, anterior margin much raised; elytra with narrow, scarcely punctured striae, interstices broad, flat, with granules toward apex. L. 1½ l. Not uncommon. *C. verrucatus*, Gyll.

B b. Antennae rust-red; elytra without white spot at base of suture.

Pitch-black; antennae and tarsi rust-red; upper-side scantily covered with brown and gray scales, sides of thorax and of elytra, apex of latter and under-side of body closely covered with grayish scales. Rostrum elongate, black, curved; thorax with central furrow and strong lateral tubercles; elytra with distinct black tubercles at sides and before apex. L. 1¾ l. Common.

C. pollinarius, Forst.

c c. Upper-side evenly and closely covered with gray scales.

Ovate. Black; antennae, tibiae and tarsi yellow-brown; body closely covered with gray scales. Rostrum not very thin; thorax scarcely broader than long, somewhat conical, very slightly constricted toward apex, closely and finely punctured, with a fine central furrow, and a small, acute tubercle on each side ; elytra with fine punctured striae, interstices not very convex, without tubercles before apex. L. 1¾ l. Very rare. *C. angulosus*, Boh.

d d. Upper-side almost bare.

Ovate. Black ; base of antennae and tarsi yellow-brown ; upper-side with scarcely any, under-side diffusely covered with gray scales. Thorax broadly and deeply constricted toward apex, with anterior margin rather strongly raised, with feeble central furrow and small lateral tubercles ; elytra with moderately strong punctured striae, interstices flat, apex spined. L. 1⅓ l. Rare. *C. picitarsis*, Gyll.

B B. Central furrow on thorax deep.

a a. Upper-side scantily covered with gray scales ; body black.

A a. Under-side closely covered with grayish scales.

Black, somewhat shiny ; under-side closely, upper-side scantily covered with grayish scales. Thorax constricted toward apex, strongly punctured, with deep central furrow and small lateral tubercles ; elytra with deep striae, interstices flat, wrinkled, with small tubercles toward apex. Male with penultimate ventral segment of abdomen with two small tubercles, placed close together, last segment concave in middle, the concavity bounded by a slight ridge. L. 1⅓ l. Common. *C. sulcicollis*, Gyll.

B b. Under-side very scantily covered with grayish scales.

Allied to *C. sulcicollis* but with tarsi pitchy-red. Male with penultimate ventral segment of abdomen simple and concavity of last segment bounded by a conical tubercle. L. 1⅓ l. Not common. *C. alliariae*, Bris.

b b. Upper-side rather closely covered with gray scales ; body leaden-black.

Leaden-black ; upper-side rather closely, under-side closely covered with gray scales. Thorax nearly as long as broad, with anterior margin not much raised, less strongly punctured than in *C. sulcicollis*, with deep central furrow and small obtuse lateral tubercles ; elytra with distinct punctured striae, interstices flat, finely wrinkled, with small tubercles before apex. L. 1—1¼ l. Scarce.

C. rapae, Gyll.

2. Ground colour of elytra blue or black-blue.

A A. Interstices on elytra flat.

a a. Elytra with short erect bristles and without white spot at base of suture.

A a. Elytra blue ; length 1⅓—1⅔ lines.

Black ; elytra blue ; under-side not closely covered with grayish scales. Thorax strongly constricted toward apex, anterior margin raised, disc coarsely punctured, with rather deep central furrow and small lateral tubercles; elytra with punctured striae; interstices flat, with a regular row of punctures, spined before apex. Not uncommon.

C. cyanipennis, Germ.

B b. Elytra black-blue ; length ⅔ line.

Black ; elytra black-blue ; under-side with gray scales. Thorax constricted toward apex, deeply punctured, with central furrow and lateral tubercles ; elytra with narrow punctured striae, interstices flat, spined toward apex. Not uncommon. *C. hirtulus*, Gyll.

b b. Elytra without bristles and with white spot at base of suture.

Short ovate, convex. Black ; elytra blue, suture black, with white spot at base ; apical margin of under-side of thorax reddish; under-side (especially at shoulders) closely covered with white scales. Thorax broader than long, slightly constricted toward apex, anterior margin slightly raised, strongly punctured, with a central furrow and somewhat pointed lateral tubercles ; elytra with distinct, obsoletely punctured striae, interstices flat, finely wrinkled. L. 1 l. Rare. *C. suturellus*, Gyll.

B B. Interstices on elytra convex.

Short ovate, rather convex. Black, somewhat shiny ; elytra blue ; under-side closely covered with grayish scales. Thorax transverse, constricted toward apex, deeply punctured, with central furrow and lateral tubercles; elytra with punctured striae ; interstices convex, wrinkled, with recumbent gray hairs and erect blackish bristles, with tubercles before apex. L. $\frac{3}{4}$ l. Rather common.

C. chalybaeus, Germ.

Coeliodes.

A. Elytra without tubercles.

a. Scutellum small, but distinct; elytra red-brown or yellow-brown.

I. Femora not toothed.

1. Elytra with fine punctured striae, interstices broad and flat.

A A. Femora without tuft of hairs.

Brown or yellow-brown; rostrum, disc of thorax and breast dark brown; with gray scales, placed more thickly on sides of thorax, breast and three more or less distinct waved bands on elytra. Sides of thorax rounded, without tubercles; outer margin of tibiae fringed before apex. L. 1—1¼ l. Common. *C. quercus*, Fab.

B B. Femora with a small toothlike tuft of white hairs.

Blackish or brown, raised anterior margin of thorax and the elytra lighter; apex of rostrum red, covered with gray scales, closely beneath, in bands (waved on elytra) above. Outer margin of tibiae toothed and fringed before apex. L. 1⅓ l. Not uncommon. *C. ruber*, Marsh.

2. Elytra with deep punctured striae, interstices only a little broader than striae and distinctly convex.

Convex. Black or dark brown; elytra and legs red-brown, with suture of former darker; with gray scales, spot at base of suture of elytra with white scales. L. 1 l. Moderately common. *C. rubicundus*, Payk.

II. Femora toothed (at least posterior pair).

Colour as in *C. quercus*, under-side much more closely covered with scales, elytra more convex, with bands narrow and sharply defined, the first united with that on other elytron and produced at suture to base, the second not much waved, the third near apex less distinct. Thorax short, with small lateral tubercles; elytra with fine punctured striae, interstices broad and flat; outer margin of tibiae with only a few spines before apex. L. 1⅓—1½ l. Not common. *C. subrufus*, Herbst.

b. Scutellum scarcely visible; elytra black.

Short ovate. Black; tibiae and tarsi dark reddish; covered with whitish scales, closely beneath, more scantily and in spots above, elytra with a composite lateral spot (rarely absent) and base and apex of suture closely covered with white scales. Thorax strongly punctured, with a shallow central furrow (deeper before and behind) and small lateral tubercles; elytra short, convex, with punctured furrows; femora toothed. L. 1—1¼ l. Common.

C. quadrimaculatus, Lin.

B. Elytra with many small tubercles placed together before apex.

Ovate. Black, dull; under-side covered closely with white, upper-side scantily with brown scales, base of suture of elytra with a velvety black spot, marked in front with a white point (sometimes absent). Forehead flat, not impressed; thorax somewhat narrowed in front, with shallow central furrow (interrupted in middle), and conical lateral tubercles; elytra with fine, indistinctly notched striae, interstices flat and broad; femora toothed. L. 1½ l. Common.

C. fuliginosus, Marsh.

C. Elytra with a regular row of small tubercles on interstices.

a. Thorax with anterior margin not raised, transverse impression near apex feeble.

Black; under-side covered with white scales. Thorax with very small lateral tubercles; elytra with deep punctured striae; femora toothed. L. 1 l. Moderately common.

C. geranii, Payk.

b. Thorax with anterior margin much raised, transverse impression near apex strong.

Similar to *C. geranii* but deeper black, shiny, with thorax more finely punctured, tubercles on elytra larger and bearing longer bristles, and femora more distinctly toothed. L. 1 l. Not uncommon. *C. exiguus*, Ol.

Mononychus.

Short ovate, rather flat. Black; funiculus of antennae reddish-yellow-brown; upper-side with hairlike light gray scales, a spot at base of suture of elytra and under-side of body closely covered with whitish scales. Thorax narrowed in front, closely punctured in wrinkles, with broad central furrow; elytra with indistinctly punctured striae, interstices flat. L. 2 l. Rather common. *M. pseudacori*, Fab.

RAMPHIDES.

Rostrum moderately robust; funiculus of antennae with first joint similar to scape, second nearly as thick but much shorter, the other five joints slender; thorax transverse; elytra convex, parallel-sided, a little wider than thorax; posterior legs stronger than front pairs, tarsi short and very slender, claws very small. *Ramphus*, Clairv.

Ramphus.

Oblong oval. Black, not very shiny, antennae (except club) yellow. Thorax narrowed in front, coarsely punctured; elytra with deep punctured striae. L. $\frac{1}{2}$ l. Common.
R. flavicornis, Clairv.

CRYPTORHYNCHIDES.

A. Channel for reception of rostrum bounded behind by mesosternum.

a. Episterna of metathorax more or less broad.

Rostrum entire at apex, scrobes commencing in or a little beyond middle, slightly oblique; thorax more or less prominent in front, with ocular lobes; mesosternum very long; intermediate abdominal segments equal, separated from first by a straight suture; legs moderately long, femora gradually clubbed, tibiae straight or nearly so, hooked at apex, with a line of stiff hairs on outer-side near apex.
Cryptorhynchus, Ill.

b. Episterna of metathorax very narrow.

Rostrum rather long, scrobes commencing in or near middle, straight; antennae moderately long, funiculus seven-jointed, with only two joints elongate; front of thorax moderately prominent, with feeble ocular lobes; scutellum absent; elytra oval, not wider than thorax; second abdominal segment longer than either third or fourth, separated from first by a slightly curved suture ; intercoxal process broad, angular in front : femora gradually clubbed, tibiae straight, strongly hooked at apex. *Acalles*, Schoenh.

B. Channel for reception of rostrum not reaching beyond anterior coxae.

Rostrum elongate, scrobes commencing before middle, very oblique, wholly visible laterally ; front of thorax straight, without ocular lobes ; prosternum effaced before anterior coxae ; intermediate abdominal segments nearly equal, separated from first by a straight suture ; femora gradually clubbed, tibiae straight, not hooked at apex.

Orobitis, Germ.

Cryptorhynchus.

Black, dull ; sides of thorax, a basal band and apical third-part of elytra closely covered with white scales ; thorax and elytra with tufts of upright black scales. Femora of male distinctly two-toothed. L. 3½—4 l. Common. *C. lapathi*, Lin.

Acalles.

A. Upper-side without bristles.

a. Elytra with an obtuse tubercle at base of second and fourth interstices.

Oblong-ovate. Pitch-black ; rostrum and antennae redbrown ; covered unequally with grayish scales. Thorax broadly and abruptly constricted in front, closely punctured in wrinkles, with central furrow behind, sides straight ; elytra with deep punctured striae, interstices ridged, second and fourth with obtuse tubercle at base. L. 1⅓ l. Moderately common. *A. roboris*, Curt.

b. Elytra without tubercles.

Ovate. Pitch-black ; rostrum, antennae and legs reddish ; scantily covered with gray scales, collected into more or less distinct transverse bands on elytra. Rostrum slightly curved, punctured ; thorax almost longer than broad, slightly narrowed in front, neither closely nor deeply punctured, without central furrow ; elytra with deep punctured striae, interstices convex. L. 1 l. Moderately common. *A. ptinoides*, Marsh.

B. Upper-side with long upright bristles (besides scales).

Ovate. Pitch-brown; rostrum, antennae and legs red-brown, with gray scales and on upper-side erect black bristles, thorax variegated with brown, elytra with white. Rostrum nearly straight; thorax somewhat longer than broad, constricted in front, closely but not deeply punctured, without central furrow; elytra convex, with moderately strong punctured striae, interstices slightly convex. L. $\frac{3}{4}$—1 l. Rather common. *A. turbatus*, Boh.

Orobitis.

Head and thorax black; elytra black-blue, often with apex reddish; antennae and legs pitch-brown; upper-side (except scutellum) bare, under-side and scutellum closely covered with white scales. Elytra with fine, scarcely punctured striae, interstices broad and flat; femora long, nearly cylindrical, not toothed. L. 1 l. Not uncommon.

O. cyaneus, Lin.

GYMNETRIDES.

A. Anterior coxae placed apart; tarsal claws free.

Prosternum with a shallow channel for reception of rostrum; other parts as in *Gymnctron*. *Miarus*, Steph.

B. Anterior coxae contiguous; tarsal claws soldered at base.

Rostrum slightly curved, scrobes commencing near middle and reaching eyes; funiculus of antennae with first two joints elongate, the other three very short; femora clubbed, tibiae more or less hooked at apex. *Gymnctron*, Schoenh.

Miarus.

A. Posterior femora toothed; thorax much broader than long.

Short ovate. Black; covered with fine recumbent hairs, on under-side whitish-gray, on upper-side yellowish-gray. Head and thorax very closely, evenly punctured, latter much broader than long; elytra scarcely longer than together broad, with punctured striae, interstices flat, each with a double row of hairs; posterior femora toothed; third joint of antennae three times as long as fourth. L. $1\frac{1}{2}$—$1\frac{3}{4}$ l. Rather common. *M. graminis*, Gyll.

B. Posterior femora simple, or posterior femora toothed and thorax nearly as long as broad.

a. Femora not toothed.

I. Third joint of antennae twice as long as fourth; pygidium of female impressed.

Nearly ovate. Black; rather closely covered with short, recumbent whitish-gray hairs, on elytra placed in rows. Head and thorax very closely and finely punctured, latter broader than long, and rather narrower than elytra, which have punctured striae and flat interstices punctured in wrinkles. Male with two tubercles, directed backward, on last ventral abdominal segment. L. 1¼—1½ l. Rather common.   *M. campanulae*, Lin.

II. Third joint of antennae scarcely half as long again as fourth; pygidium of female not impressed.

Black, scantily covered with erect gray hairs, not placed in rows on elytra. Thorax closely punctured, sides rounded behind; scutellum triangular, pointed, closely covered with gray-white scales; elytra with punctured striae, sides tolerably straight; apex of abdomen simple. L. 1 l. Rare.
M. micros, Germ.

b. Posterior femora toothed.

Oblong ovate. Black, with gray recumbent hairs. Head and thorax finely and closely punctured, latter nearly as long as broad; elytra one-third longer than together broad, with strong punctured striae, interstices narrow, with a regular row of gray-white hairs; posterior femora with a very small tooth. L. 1⅓ l. Rather common.
M. plantarum, Dej.

Gymnetron.

A. Apex of elytra jointly rounded, leaving at most apex of pygidium uncovered.

a. At least anterior femora with a little tooth.

Black; base of antennae, tibiae and tarsi rust-red; elytra with or without a red spot, variable in size; body covered with fine gray hairs. Thorax not much broader than long, narrowed in front, sides almost straight, base bisinuate, very finely and closely punctured; elytra with deep punctured striae, interstices with a row of erect, white bristles. L. ¾ l. Common.   *G. pascuorum*, Gyll.

b. Femora not toothed.

I. Sides of thorax and the breast closely covered with whitish-gray or yellowish-white scales ; length 1¼—1½ lines.

A A. Base of thorax bisinuate.

Black ; elytra and legs dark rust-red ; closely covered throughout (less so on disc of thorax than at sides) with white-gray scales. Sides of thorax rounded ; elytra with indistinct punctured striae. L. 1½ l. Not uncommon.

G. villosulus, Gyll.

B B. Base of thorax nearly straight.

Black ; elytra (except suture and side margin), base of antennae and legs red-brown ; body covered with fine gray hairs, sides of thorax (nearly to middle) and breast with yellowish-white scales. Elytra with indistinctly punctured striae. L. 1¼ l. Moderately common.

G. beccabungae, Lin.

II. Sides of thorax (with whole upper and under sides) covered with fine gray hairs ; length ⅔—⅘ lines.

1. Elytra black, with red markings.

Black ; elytra with two oblique red transverse bands, interrupted at suture ; base of antennae, tibiae and tarsi rust-red. Elytra with punctured striae, hairs on interstices in rows. L. ¾—⅘ l. Rather common. *G. labilis,* Herbst.

2. Elytra unicolorous black.

A A. Tibiae and tarsi red.

Oblong-ovate, convex. Black, scantily covered with gray hairs ; base of antennae, tibiae and tarsi red. Rostrum parallel-sided, not much curved ; thorax finely punctured ; elytra with deep punctured striae, suture and interstices with rows of bristles. L. ¾ l. Rare.

G. rostellum, Herbst.

B B. Tibiae and tarsi black.

Similar to *G. rostellum* but with bristles on elytra shorter and more diffusely placed ; rostrum shorter. L. ⅔ l. Not very common. *G. melanarius,* Germ.

B. Elytra separately rounded at apex, leaving pygidium wholly uncovered.

a. Thorax almost as long as broad.

Ovate. Black, scantily covered with gray hairs. Rostrum not, or scarcely longer than thorax, narrowed toward apex, punctured in front, smooth behind, with feeble central furrow ; thorax rounded at base, narrowed in front, very closely punctured ; elytra somewhat broader than thorax, with distinct punctured striae ; femora usually with a small tooth. L. 1¼—1½ l.   Common.   *G. noctis,* Herbst.

b. Thorax much broader than long.

I. Femora toothed.

Ovate. Black, scantily covered with rather long, recumbent gray hairs. Rostrum round, moderately long, not much curved ; thorax indistinctly ridged in middle ; elytra with punctured striae ; femora with a sharp tooth (except anterior pair in female). Not common. L. 1⅓ l.

G. collinus, Gyll.

II. Femora not toothed.

Ovate. Black, somewhat shiny ; scantily covered with short, recumbent whitish-gray hairs. Rostrum scarcely narrowed toward apex, curved, in female nearly impunctate at apex ; elytra with distinct punctured striae ; legs strong, femora much thickened. L. 1½ l.   Not common.

G. linariae, Panz.

CIONIDES.

A. Intercoxal process moderately broad, triangular.

Club of antennae composed of three joints, sometimes loosely, at others closely attached ; eyes rounded ; scutellum scarcely visible, or absent ; elytra not broader than thorax ; legs moderately long, femora sometimes toothed, tibiae slender, straight, tarsal claws soldered or free ; second abdominal segment larger than third and fourth together, separated from first by a nearly straight suture.

Nanophyes, Schoenh.

B. Intercoxal process very broad, parallel-sided and truncate.

Club of antennae distinctly articulate ; eyes oblong ; scutellum rather large ; elytra distinctly broader than thorax ; legs short, femora toothed (except often in female), tibiae more or less curved at base, tarsal claws soldered ; second abdominal segment larger than third and fourth together, separated from first by a straight suture.

Cionus, Clairv.

Nanophyes.

A. Femora not toothed.

Black, shiny; base of antennae and legs reddish-yellow; elytra with an abbreviated band and behind this a spot yellow-brown and covered with whitish-gray hairs, rest of body with fine gray hairs, thickly matted on breast. Colour variable, thorax, elytra (except basal spot) and abdomen being sometimes red-yellow or yellow-brown. L. ⅔ l. Common. *N. lythri*, Fab.

B. Femora with two small sharp spines on under-side.

Black, shiny; base of antennae, elytra (except a common, triangular spot at base and the outer margin) and legs red-yellow, apex of femora black. Elytra less pointed behind than in *N. lythri* with interstices flat, legs, antennae and rostrum thinner, pubescence less distinct. L. ½ l. Rare. *N. gracilis*, Redt.

Cionus.

A. Prosternum excavated before coxae and excised at apex; alternate interstices on elytra with rows of larger punctures; third joint of antennae as long as second.

   a. Thorax without brown basal spot.

      I. Interstices on elytra checkered.

         1. The two black velvety sutural spots united by a pale spot; predominant colour of upper-side black.

         A A. Thorax covered throughout with grayish or yellowish-white hairs.

Black; thorax, a spot at shoulder, breast and legs covered with grayish or yellowish-white hairs; elytra brown, alternate interstices raised, checkered with black and white, suture with two velvety black spots, united by a pale spot. Club of antennae oblong oval. L. 2—2¼ l. Common.

C. scrophulariae, Lin.

         B B. Only sides of thorax covered with yellowish hairs.

Black; antennae and tarsi yellow-red; sides of thorax

and of breast and a spot at shoulder covered with yellowish hairs; legs scantily covered with gray hairs; elytra with alternate interstices raised, checkered with black and gray, suture with two velvety black spots, united by a pale spot. Club of antennae spindle-shaped.   L. 1¾ l.   Common.

C. verbasci, Fab.

2. The two black velvety sutural spots not united by a pale spot.

A A. Hinder sutural spot smaller than anterior one.

Red-brown; covered rather closely throughout with greenish-gray hairs; elytra with alternate interstices slightly raised, checkered with badly defined pale and brown naked spots, suture with two round, simple, velvety black spots, the hinder one smaller. Rostrum of female punctured in wrinkles and not narrowed toward apex. L. 2—2½ l.   Not common.   C. thapsus, Fab.

B B. Sutural spots equal in size.

Similar to C. thapsus but with sutural spots on elytra equal in size; rostrum of female smooth and narrowed toward apex.   L. 1¾ l.   Common.   C. hortulanus, Marsh.

II. Upper-side evenly covered with greenish-white or greenish-gray matted hairs and also with erect white bristles.

Brown; with matted hairs as above; antennae and legs reddish-yellow-brown; suture of elytra with a small velvety black spot somewhat before middle and a still smaller one (sometimes absent) before apex.   L. 1½—2 l.   Rare.

C. olens, Fab.

b. Thorax with brown basal spot.

Brown; upper-side closely covered with grayish-white matted hairs, thorax with a large brown spot at base, elytra with some interstices indistinctly raised, checkered with brown and white, suture with a large quadrangular spot (composed of small brown and black spots) on anterior half and a large round velvety black spot, bordered with white, before apex; sides of thorax and breast thickly covered with yellowish-white matted hairs.   L. 1¼—1¾ l. Rather common.   C. blattariae, Fab.

B. Prosternum not excavated before coxae, entire in front; alternate interstices on elytra without rows of larger

punctures ; third joint of antennae shorter than second.
Reddish-brown ; with scanty gray pubescence ; elytra
with alternate interstices somewhat raised, checkered with
spots of black and white hairs, suture with three velvety-
black spots, middle one largest. Thorax very short, sides
rounded. L. 1¼ l. Rather common. *C. pulchellus*, Herbst.

TYCHIIDES.

A. Second abdominal segment enclosing third at sides
and touching fourth.

  a. Elytra separately rounded at apex, leaving pygidium
at least partly uncovered.

Funiculus of antennae with six joints ; in other respects
as in *Tychius*.   *Sibynes*, Schoenh.

  b. Elytra jointly rounded at apex, covering pygidium.

Antennae moderately long, not very robust, funiculus
generally with seven, more rarely with six joints, club oval
or oblong oval, articulate; eyes round or short oval ; second
abdominal segment a little longer than either third or
fourth, suture separating it from first, either straight or
slightly angular in middle ; episterna of metathorax not
very narrow.   *Tychius*, Schoenh.

B. Second abdominal segment leaving third free.

Antennae rather short, slender, funiculus with seven
joints, first much longer than second, club short oval,
nearly solid ; rostrum rather shorter and thicker than in
Tychius ; eyes oval ; elytra nearly covering pygidium ;
second abdominal segment nearly as long as third and
fourth together, separated from first by a straight suture ;
episterna of metathorax rather narrow.   *Elleschus*, Steph.

Sibynes.

A. Base of thorax bisinuate ; elytra nearly oblong-quad-
rangular, rather flat.

  a. Elytra unicolorous or with lighter markings.

    I. Upper-side covered with gray or yellowish hair-
like scales.

Ovate. Black ; under-side covered with white scales ;

upper-side as above, central line and sides of thorax, suture and some interstices of elytra often rather lighter coloured. Scutellum rounded, raised. L. 1½ l. Rare. *S. canus*, Herbst.

II. Upper-side with reddish-brown matted hairs.

Ovate, convex. Black; under-side with white scales; upper-side as above, thorax with three indistinct whitish lines, each interstice on elytra with a line of white scales. Thorax broad, deeply constricted toward apex; elytra with fine punctured striae, interstices flat. L. 1⅓ l. Not common.   *S. potentillae*, Germ.

b. Elytra with a common dark spot at suture.

I. Upper-side with grayish scales; length 1½ line.

Oblong-ovate. Black; base of antennae, tibiae and tarsi red-brown; covered with white scales, thorax with a large dorsal spot, somewhat narrowed in front, light brown; elytra with a common spot on anterior half of suture light brown. Thorax nearly as long as broad, constricted toward apex; elytra with indistinct punctured striae. Not uncommon.   *S. arenariae*, Steph.

II. Upper-side with yellowish-red scales; length 1—1¼ line.

Black; base of antennae, tibiae and tarsi reddish-brown; under-side with white, upper-side with grayish-yellow scales, two broad longitudinal lines on thorax and an oblong spot on anterior half of suture of elytra brown. Not very common.   *S. primitus*, Herbst.

B. Base of thorax tolerably straight; elytra oblong-ovate; somewhat convex.

Oblong-ovate. Black; rostrum, antennae and legs pitch-brown; upper-side covered with brown, under-side with white scales. Thorax abruptly narrowed in front, constricted toward apex, closely and finely punctured; elytra with fine punctured striae, interstices flat, with rows of short white bristles. L. 1—1¼ l. Rare. *S. sodalis*, Germ.

Tychius.

A. Funiculus of antennae with seven joints.

   a. Length 1—1⅔ lines.

    I. Rostrum more or less narrowed toward apex.

1. Posterior femora with a strong, sharp tooth.

Black ; antennae and tibiae red-brown ; under-side covered with white scales, upper-side with coppery or brassy ones, central line of thorax, suture and two spots on each elytron (composed of short lines) white. L. 1⅓—1⅔ l. Not common. *T. quinquepunctatus*, Lin.

2. Posterior femora very obtusely toothed.

A A. Thorax much narrower than elytra, only slightly narrowed behind and scarcely rounded at sides.

Ovate. Black ; apex of rostrum, antennae, tibiae and tarsi rust-red ; under-side with white scales, upper-side with gray and brown hairlike scales, central line and sides of thorax, suture and a broad longitudinal streak (formed of several confluent lines) at sides of elytra white. L. 1½—1¾ l. Rather common. *T. venustus*, Fab.

B B. Thorax a little narrower than elytra, more or less narrowed behind, sides distinctly rounded.

a a. Scales narrow, hairlike.

Elongate ovate. Black ; apex of rostrum, antennae, tibiae and tarsi rust-red; under-side with white, upper-side with hairlike brown scales, central line of thorax, suture and alternate interstices on elytra whitish. L. 1⅓ l. Rare. *T. polylineatus*, Germ.

b b. Scales roundish.

Ovate. Pitch-brown; greater part of rostrum, the antennae and legs reddish-yellow; under-side with chalky-white, upper-side with straw-coloured rounded scales. L. 1¼ l. Moderately common. *T. squamulatus*, Gyll.

3. Femora not toothed.

A A. Antennae red-yellow at base, black at apex.

a a. Rostrum gradually and slightly narrowed toward apex.

Elliptic, not very convex. Black or brown ; base of antennae, tibiae, tarsi, and often apex of rostrum red ; under-side closely covered with white, upper-side with hairlike, whitish-gray or yellowish scales, placed more closely and rather paler on central line of thorax and suture of elytra. L. 1¼ l. Not very common.

T. Schneideri, Herbst.

b b. Rostrum subulate.

Black, with grayish pubescence ; thorax with central line and elytra with suture white ; base of antennae, tibiae and tarsi red. Alternate interstices on elytra rarely paler. L. 1¼ l. Not very common.   *T. lineatulus*, Steph. (Bris.)

B B. Antennae entirely red.

a a. Femora red.

A a. Thorax without white central line ; scales hairlike.

Oblong, convex. Pitch-black ; apex of rostrum, the antennae and legs rust-red ; under-side covered with whitish scales, upper-side with gray or yellowish-gray recumbent hairs, suture of elytra usually lighter. Rostrum scarcely as long as head and thorax together, moderately curved, somewhat narrowed at apex. Anterior tibiae of male somewhat curved, of female straight. L. 1 l. Rather common.   *T. tomentosus*, Herbst.

B b. Thorax with white central line; scales oval oblong.

Short ovate. Pitch-black, convex ; rostrum, antennae and legs red-yellow ; upper-side closely covered with yellowish oval oblong scales, under-side with whitish oval scales. Rostrum narrowed at apex ; thorax rounded at sides ; elytra very short oval, with fine punctured striae. L. 1 l. Moderately common.   *T. curtus*, Bris.

b b. Femora black.   .

Similar to *T. tomentosus* but with apex of rostrum narrower, femora black and anterior tibiae of male with a strong sharp tooth in middle. L. 1 l. Moderately common.   *T. meliloti*, Steph.

II. Rostrum not narrowed toward apex.

1. Femora not toothed.

Oblong, convex. Black ; antennae, tibiae and tarsi rust-red ; upper-side covered with gray, under-side with white pubescence, base of thorax thinly margined with white. Similar to *T. tomentosus* but with rostrum longer, not narrowed toward apex, thorax narrower and elytra with more distinct punctured striae. L. 1 l. Not common.

T. tibialis, Boh.

2. Femora toothed.

Oblong-ovate. Pitch-black ; head, rostrum, antennae and legs rust-red; head almost bare, thorax covered with gray pubescence, scantily on disc, more closely on a curved spot on each side, elytra closely and unequally covered with grayish hairlike scales, under-side with white scales ; scutellum black. Thorax a little broader than long ; elytra with fine, scarcely punctured striae. L. 1⅓ l. Rare.

T. haematocephalus, Gyll.

b. Length ¾ line.

Black ; base of antennae, tibiae and tarsi reddish ; covered with gray hairlike scales. Rostrum gradually and slightly narrowed toward apex ; thorax moderately rounded at sides, slightly narrowed at base. Not uncommon.

T. pygmaeus, Bris.

B. Funiculus of antennae with six joints.

Oblong, convex. Black ; base of antennae, tibiae and tarsi rust-red ; under-side covered with white scales, upper-side more scantily with gray pubescence. Rostrum as long as head and thorax, gradually and slightly narrowed toward apex, not much curved ; elytra with distinct punctured striae. L. ⅔—1 l. Common.

T. picirostris, Fab.

Elleschus.

A. Thorax reddish-yellow-brown.

Reddish-yellow-brown, with gray pubescence ; head, breast, base of abdomen and a large oblong common spot on front part of elytra blackish ; suture and several lines at base of elytra with pubescence more closely placed and lighter. Elytra with punctured striae. L. 1¼—1½ l. Rare.

E. scanicus, Payk.

B. Thorax black.

Black, closely covered with gray pubescence ; antennae and generally tibiae and tarsi yellow-brown. Elytra with deep punctured striae, middle of each near suture with a small, less closely pubescent and therefore darker spot.

L. 1¼ l. Common.

E. bipunctatus, Lin.

ANTHONOMIDES.

A. Posterior legs formed for leaping; eyes much approximated above; generally contiguous.

Rostrum more or less long, not very robust, folded beneath when at rest; funiculus of antennae generally with six, more rarely with seven joints, first three joints elongate; elytra much broader than thorax and jointly rounded at apex, leaving pygidium more or less (generally only slightly) uncovered; prosternum very short; tarsal claws appendiculate.   *Orchestes,* Ill.

B. Posterior legs not formed for leaping; eyes placed apart.

a. Prosternum not emarginate and not very short; tarsal claws simple.

Rostrum elongate, not very robust; funiculus of antennae with seven joints, first and second elongate, former much longer than latter; elytra little broader than thorax, separately rounded at apex, imperfectly covering pygidium.   *Acalyptus,* Schoenh.

b. Prosternum widely emarginate, very short; tarsal claws appendiculate or bifid.

Rostrum more or less long, slender; funiculus of antennae with seven joints, first two elongate; elytra a little broader than thorax, almost entirely covering abdomen; anterior legs larger than the others. *Anthonomus,* Germ.

Orchestes.

A. Funiculus of antennae with six joints.

a. Posterior femora toothed.

I. Thorax red or yellow-brown.

1. Elytra unicolorous; head reddish.

A A. Abdomen red-yellow; posterior femora toothed sawlike.

Reddish-yellow-brown; eyes and breast black; closely covered with gray hairs, generally placed more thickly on a large spot (pointed behind) on base of elytra. L. $1\frac{1}{2}$— $1\frac{3}{4}$ l. Common.   *O. quercus,* Lin.

B B. Abdomen black; posterior femora with only one distinct tooth.

Upper-side rather darker than in *O. quercus*; eyes, apex of rostrum and under-side (except of head) black; scutellum closely covered with white hairs. L. 1⅓—1½ l. Moderately common. *O. scutellaris*, Gyll.

2. Elytra red, with a small round spot in middle of base and a large common spot somewhat behind middle black; head black.

Black; thorax, apex of abdomen and tarsi yellow-red, elytra as above; pubescence gray. Posterior femora finely sawlike beneath. L. 1⅓ l. Common. *O. alni*, Lin.

In variety *ferrugineus*, Marsh., the elytra are unicolorous red.

II. Thorax black.

1. Elytra without white spot at base of suture.

A A. Pubescence on elytra arranged in markings.

a a. Elytra closely checkered.

Oblong oval. Black; antennae and tarsi red-yellow; closely checkered with whitish, gray and reddish hairs. Posterior femora large, sawlike beneath. L. 1½ l. Moderately common. *O. ilicis*, Fab.

b b. Elytra with indistinct white bands and a quadrangular common red-brown spot behind scutellum.

Nearly ovate. Black; antennae and tarsi reddish-yellow; under-side with scanty gray pubescence; rostrum with a bare, shiny central line, thorax with yellow-gray pubescence and sprinkled with longer black hairs, elytra with black recumbent hairs, indistinctly banded with white pubescence, with a quadrangular common red-brown spot behind scutellum. Posterior femora with a row of indistinct teeth beneath. L. 1¼ l. Rare. *O. sparsus*, Gyll.

B B. Pubescence on elytra uniform.

a a. Striae on elytra distinctly punctured, length 1¼ lines.

Elongate oval. Black, antennae and tarsi light yellow-brown; thinly covered throughout with fine gray pubescence.   All femora with one tooth before apex.   Common.

O. fagi, Lin.

b b. Striae on elytra indistinctly punctured ;
length ⁴⁄₅—1 line.

Oblong oval.   Black ; antennae and tarsi reddish-brown ;
closely covered with whitish-gray pubescence.   Posterior
femora angularly dilated beneath.   Moderately common.

O. pratensis, Germ.

2. Elytra with white spot at base of suture.

Ovate.   Black ; antennae and tarsi brown ; covered with
long, rough hairs, base of suture of elytra with white hair.
Thorax short, with a central furrow behind ; femora angularly dilated with a row of teeth.   L. 1⅓ l.   Not common.

O. iota, Fab.

b. Posterior femora not toothed.

I. Thorax reddish-yellow.

Oval.   Reddish-yellow ; eyes, breast, abdomen and a
ring before apex of posterior femora black ; elytra with a
small spot at shoulder and a broad toothed band in middle
black-brown.   L. 1¼ l.   Rare.   *O. lonicerae*, Herbst.

II. Thorax black.

1. Tibiae black.

Oval.   Black ; antennae and tarsi reddish-yellow ; elytra
with base of suture and two transverse bands (composed of
small spots) in middle covered with grayish pubescence.
Thorax strongly punctured ; elytra with strong punctured
striae, interstices nearly impunctate, shiny.   L. 1¼ l.
Moderately common.   *O. rusci*, Herbst.

2. Tibiae reddish-yellow.

Oval.   Black ; elytra with a large common heart-shaped
spot at base and a transverse band before middle covered
with white or yellow hairs ; antennae, tibiae and tarsi
reddish-yellow ; thorax covered with white or yellow hairs.
L. 1¼ l.   Moderately common.   *O. avellanae*, Don.

B. Funiculus of antennae with seven joints.

a. Elytra unicolorous black.

I. Antennae and legs black ; length 1⅓ lines.

Oblong-ovate.   Black ; with fine gray pubescence, placed

more closely and lighter coloured on scutellum and breast. Thorax oblong, conical ; elytra with deep punctured striae. Common. *O. stigma*, Germ.

II. Antennae and legs partly red-yellow ; length $\frac{3}{4}$ line.

Oblong. Black ; base of antennae and tibiae red-yellow ; pubescence scanty, gray, on breast closer, whitish. Elytra convex, with deep punctured striae. Rather common. *O. saliceti*, Fab.

b. Elytra with lighter markings.

Oval. Black ; base of antennae red ; elytra with two curved bands of white hairs, the anterior one dilated at suture into a yellowish spot. Elytra with punctured striae. L. $1\frac{1}{4}$ l. Common. *O. salicis*, Lin.

Acalyptus.

Ovate, rather flat. Black ; antennae and legs reddish-yellow ; elytra reddish-yellow-brown, with base, suture and side margin blackish ; pubescence close, gray, shiny. Sides of thorax scarcely rounded ; elytra with punctured striae. L. 1 l. Not common. *A. rufipennis*, Gyll.

Anthonomus.

A. Elytra with a transverse or oblique band of whitish hairs ; third stria joined behind with sixth.

a. White band on elytra transverse.

I. White band on elytra distinct.

1. Thorax and elytra not very convex, sides of former only slightly rounded.

A A. Tooth on anterior femora strong.

Brownish-red, with scanty whitish pubescence ; head, rostrum and breast pitch-black ; scutellum and a broad transverse band behind middle of elytra closely covered with whitish hairs. Rostrum elongate, somewhat curved, dull ; scutellum elongate-oval ; elytra diffusely covered with pale hairs, with a bare band in middle ; anterior femora strongly, posterior pair feebly toothed. L. $1\frac{1}{2}$ l. Common. *A. ulmi*, De G.

B B. Tooth on anterior femora only moderately strong.

Black, with scanty whitish pubescence ; antennae and legs rust-red ; scutellum and two bands on elytra covered with whitish hairs, latter with a broad bare band in middle. Similar to *A. ulmi*, but with rostrum shorter and thicker, antennae inserted nearer its apex, eyes less prominent, scutellum broader, thorax shorter, interstices on elytra broader, tooth of anterior femora shorter and feebler. L. 1½ l. Common. *A. pedicularius*, Lin.

2. Thorax and elytra very convex, sides of former distinctly rounded.

Ovate, strongly convex. Dark red or brown-red ; club of antennae brown ; thorax with white central line ; elytra with white bands, the anterior band curved toward scutellum. Rostrum moderately elongate, dull ; antennae hairy ; thorax very transverse, narrowed slightly at base, strongly at apex, sides distinctly rounded ; interstices of striae on elytra flat, rather shiny, with extremely fine transverse wrinkles and a row of very fine punctures ; anterior femora with a strong tooth, hinder pairs with a small obtuse tooth ; anterior tibiae dilated on inner-side. L., including rostrum, 2¼ l. Rare.   *A. Chevrolati*, Des L.

II. White band on elytra indistinct.

Ovate. Pitch-black, rostrum, antennae and legs lighter ; thorax with light central line scarcely perceptible ; elytra with whitish-gray or yellowish hairs scarcely forming true bands. Narrower than *A. ulmi*, more parallel-sided ; head depressed in front, eyes smaller and less prominent, rostrum short, thick, curved, apex somewhat attenuate beneath, not shiny ; antennae less slender, second joint shorter, base of club narrower ; legs more slender, teeth more acute. L. 1½ l. Rare.   *A. conspersus*, Des L.

b. White band on elytra oblique.

Pitch-brown, with gray pubescence; elytra rust-red, with an oblique band behind middle (dilated on outer-side) covered with whitish hairs and bordered with black ; scutellum oval, covered with white hairs ; antennae and legs rust-red. Rostrum rather long and curved; femora toothed, anterior pair strongly, intermediate pair moderately strongly, and posterior pair feebly. L. 1½—2 l. Rather common.
A. pomorum, Lin.

B. Elytra without band of whitish hairs; third stria joined behind with eighth.

a. Anterior femora with a strong tooth ; rostrum shiny.

Black ; base of antennae yellow-brown ; disc of thorax and elytra brown-red, margins of latter generally black ; scutellum white. Thorax deeply punctured ; elytra with deep punctured striae ; femora with a pointed tooth. L. 1¼—1½ l. Rare. *A. varians,* Payk.

b. Femora feebly toothed; rostrum dull.

I. Head and rostrum dark rust-red.

Rather short ovate, convex, upper-side almost bare. Dark rust-red. Eyes prominent and much approximated ; rostrum rather short and dull, dilated at apex, moderately curved ; thorax closely and finely punctured, sides only slightly rounded ; elytra scarcely curved at sides, with deep punctured striae, interstices convex, with a row of punctures ; anterior tibiae sinuate on outer-side at apex. L., including rostrum, 1¾ l. Rare. *A. britannus,* Des L.

II. Head and rostrum black.

1. Femora not strongly dilated in middle, and not very abruptly narrowed before apex.

Black, rather shiny; scape of antennae sometimes yellow-brown ; pubescence fine, gray, closer on breast; scutellum whitish. Thorax extremely closely and finely punctured, base fully half as broad again as apex ; elytra with strong punctured striae, interstices rather convex ; anterior tibiae gradually curved at base; first joint of funiculus of antennae much longer than broad. L. 1 l. Common.

A. rubi, Herbst.

2. Femora strongly dilated in middle and abruptly narrowed before apex.

Similar to *A. rubi,* but with elytra sometimes paler, anterior tibiae abruptly bent at base and thorax less narrowed at apex ; first joint of funiculus of antennae nearly as broad as long. L. ¾ l. Not common. *A. comari,* Crotch.

BALANINIDES.

Antennae long, slender, funiculus with seven joints, first three elongate, the others variable in length ; elytra sepa-

rately rounded at apex, a little broader than thorax.
Balaninus, Germ.

Balaninus.

A. Femora with a large triangular tooth before apex.

a. Rostrum entirely red or red-brown, moderately long,
thickened at base.

    I. Last joint of funiculus of antennae at least double
as long as broad, reversed conical.

Oblong-ovate. Black; antennae red-brown; rostrum
red; with grayish-yellow scales, central line and sides of
thorax lighter, elytra with some indistinct darker bands,
the hairlike scales placed on hinder half of suture more
thickly and upright; rostrum striate and punctured. L. 3 l.
Rather common.   *B. glandium*, Marsh.

    II. Last joints of funiculus of antennae scarcely
longer than broad.

Ovate. Black; legs dark rust-brown; rostrum red-
brown; with gray or yellowish scales, scutellum and
irregular spots (here and there confluent into bands) on
elytra lighter. Funiculus of antennae thickly pubescent;
rostrum striate and punctured at base, in male slightly, in
female strongly curved. L. 3—3½ l. Common.

                               B. nucum, Lin.

b. Rostrum yellow-red, with black apex, very long and
thin.

Ovate, shorter and broader than *B. glandium.* Black;
legs dark rust-brown; rostrum yellow-red, apex black;
checkered with gray or yellowish-gray scales. All joints
of funiculus of antennae elongate, only slightly thickened
at apex, scantily pubescent; rostrum much curved, indistinctly punctured at base. L. 2¼—2¾ l. Not common.

                     B. tessellatus, Fourc.

B. Femora with at most a small tooth before apex.

    a. Last joints of funiculus of antennae oblong; pygidium
uncovered; length 1¼—2 lines.

        I. Upper-side brown-red or red-brown.

            1. Under-side brown-red.

Brown-red; eyes black; thorax and under-side covered
tolerably evenly, elytra in spots with yellowish-white hair-

like scales, spots on latter forming two more or less regular bands. Thorax with fine and very close granulate punctuation; elytra with distinct punctured striae, interstices even, punctured in wrinkles; femora with a feeble tooth. L. $1\frac{1}{3}$ l. Not common. *B. cerasorum*, Herbst.

2. Under-side black.

Upper-side red-brown, under-side black; antennae rust-red; legs pale; rostrum red; elytra pale reddish-yellow; covered unequally with narrow grayish-white scales, elytra with base, suture and a more or less distinct transverse band somewhat behind middle more closely covered with whitish scales. Rostrum and antennae much shorter than in *B. cerasorum*; elytra with punctured striae, interstices flat, wrinkled; femora scarcely toothed beneath. L. $1\frac{1}{4}$ l. Not common. *B. rubidus*, Gyll.

II. Upper-side black.

Black; base of antennae and usually also apex of tibiae and tarsi red-brown; under-side covered evenly and more closely, upper-side more scantily with gray hairlike scales, scutellum and a band (abbreviated at each side) behind middle of elytra with more closely placed grayish or yellowish white scales. L. 2 l. Rather common.
B. villosus, Herbst.

b. Last joints of funiculus of antennae short; pygidium entirely or nearly entirely covered; length $\frac{2}{3}$—$\frac{4}{5}$ line.

I. Whole of breast closely covered with white scales.

Black; base of antennae reddish-yellow-brown; upper-side very scantily covered with gray pubescence, scutellum and under-side closely covered with white scales. Rostrum almost as long as body; thorax strongly and closely punctured; elytra with punctured striae, interstices flat, wrinkled. L. $\frac{2}{3}$—$\frac{3}{4}$ l. Common. *B. brassicae*, Fab.

II. Only sides of breast closely covered with white scales.

Similar to *B. brassicae* but with upper-side still more scantily pubescent, middle of breast with fine gray hairs only, antennae (except club) reddish-yellow-brown and interstices on elytra narrower. Apex of rostrum red in male. L. $\frac{3}{5}$—$\frac{4}{5}$ l. Common. *B. pyrrhoceras*, Marsh.

MAGDALINIDES.

Antennae imperfectly elbowed, scape slightly curved, funiculus with seven joints, first two elongate, club oblong oval, pointed; eyes placed moderately far apart; elytra not broader than thorax, broadly rounded at apex; tibiae more or less slender, not toothed on inner-side; second abdominal segment scarcely so long as third and fourth together, separated from first by a curved suture; body cylindrical.

Magdalinus, Schoenh.

Magdalinus.

A. Femora toothed.

a. Thorax as long or nearly as long as broad.

I. Sides of thorax not toothed.

1. Head punctured.

A A. Interstices on elytra confusedly punctured, without wrinkles; rostrum nearly straight.

Black, with bluish reflection; elytra dark blue. Head elongate, rather flat; eyes not approximated, forehead elongate; rostrum as long as thorax, closely punctured; thorax longer than broad, rather flat, constricted in front, strongly punctured; elytra broader behind, with fine punctured striae, punctures on interstices rather large, with traces here and there of irregular double rows; anterior femora with a large pointed tooth. L. 2½ l. Not common.

M. phlegmaticus, Herbst.

B B. Interstices on elytra with very fine wrinkles, besides punctures; rostrum curved.

a a. Tarsal claws with a tooth at base; length 2½ lines.

Black, rather shiny; breast marked with a little spot of whitish scales on each side. Rostrum as long as thorax, curved, somewhat thickened at apex in male; thorax narrowed in front, strongly punctured; elytra with punctured striae, the punctures in the striae oblong and separated by much raised transverse wrinkles, connected with those in the neighbouring striae across the punctured interstices; under-side distinctly punctured; anterior femora strongly toothed. L. 2½—3 l. Scarce. *M. carbonarius*, Lin.

b b. Tarsal claws simple; length 1¾—2 lines.

Black; elytra black-blue. Rostrum almost longer than thorax, strongly curved; thorax much narrowed in front, scarcely constricted, very closely punctured; elytra with deep punctured striae, interstices extremely finely shagreened, with a regular row of punctures; anterior femora with a large pointed tooth. L. 1¾—2 l. Rare.

M. duplicatus, Germ.

2. Head impunctate.

Similar to *M. duplicatus,* but more ovate, rostrum more robust, striae on elytra wider. Rare. *M. Heydeni,* Des L.

II. Sides of thorax toothed in front.

Black, dull; antennae and tarsi pitch-brown. Rostrum short, moderately curved; thorax slightly convex, closely punctured; elytra with moderately strong punctured striae, interstices flat, closely shagreened; anterior femora with a pointed tooth. L. 2½ l. Common.

M. atramentarius, Marsh.

b. Thorax much broader than long.

Black, dull. Rostrum distinctly punctured at base; thorax rounded at sides, narrowed at apex, very closely punctured; elytra with strong, somewhat distinctly punctured striae, interstices somewhat convex, very finely and extremely closely granulate; femora indistinctly toothed. Rostrum of male somewhat dilated at apex. L. 1½—1¾ l. Not uncommon. *M. cerasi,* Lin.

B. Femora not toothed.

a. Thorax with distinct lateral tubercles.

Black, dull; antennae (except club) reddish-yellow-brown. Rostrum as long as head, straight, very finely punctured; thorax narrowed in front, very closely punctured; elytra nearly cylindrical, with rather deep notched striae, interstices narrow, convex, finely wrinkled. L. 1⅓—1½ l. Common. *M. pruni,* Lin.

b. Thorax without lateral tubercles.

Black, dull, elytra more shiny; antennae (except club) rust-brown, club long. Rostrum as long as head; thorax constricted in front, rather flat, very closely punctured, with three more or less distinct depressions (two on disc and one in middle of base); elytra with deep punctured

striae, interstices convex, very finely wrinkled.   L. 1—1¼ l.
Rare.   *M. barbicornis*, Latr.

RHINOMACERIDES.

A. Head transverse, nearly quadrangular.

Maxillary palpi slender and flexible, labrum distinct;
rostrum scarcely longer than head, rather strongly dilated
in front; elytra covering pygidium; tarsal claws simple;
fifth abdominal segment as large as fourth; episterna of
metathorax narrow.   *Rhinomacer*, Fab.

B. Head longer than broad, nearly cylindrical.

Rostrum variable; elytra much broader than thorax,
nearly always leaving pygidium uncovered; intermediate
coxae placed apart, tarsi moderately long, first joint shorter
than second and third together, claws appendiculate or
toothed; fifth abdominal segment rarely as large as fourth;
episterna of metathorax more or less long.
                                   Rhynchites, Herbst.

Rhinomacer.

Oblong.   Black; antennae and legs reddish-yellow;
closely covered with gray hairs.   Punctuation close, distinct.   L. 1¾ l.   Not common.   *R. attelaboides*, Fab.

Rhynchites.

A. Posterior coxae oval, placed distinctly apart from episterna of metathorax.

a. Forehead only slightly impressed.

Blue or green, with golden reflection, without hairs.
Thorax and elytra rather more closely and finely punc
tured than in *B. populi*, former in female with a pointed
spine, directed forward, on each side in front.   L. 2½—3 l.
Rather common.   *R. betuleti*, Fab.

b. Forehead with a rather deep furrow.

Upper-side green, bronze or coppery; under-side, rostrum,
and legs blue, without hairs.   Thorax broader than long,
finely punctured, in female with a pointed spine, directed
forward, on each side in front; elytra with irregular punc-

tured striae. L. 2—2½ l. Rather common.

R. populi, Lin.

B. Posterior coxae strongly transverse, more or less prolonged along posterior margin of episterna of metathorax.

a. Head without circular furrow behind eyes.

Tibiae toothed on inner-side.

I. Elytra confusedly punctured in wrinkles, with scarcely any traces of striae.

1. Eyes rather flat ; length 3 lines.

Greenish or purple, with golden reflection, apex of rostrum, antennae and tarsi black-blue, with long hairs. Head elongate ; thorax closely and deeply wrinkled, in female with a pointed spine, directed forward, on each side in front. Rare. *R. auratus*, Scop.

2. Eyes hemispherical ; length 1¾—2 lines.

Purple, with golden reflection ; rostrum, antennae and legs black-blue ; hairs long, but not closely placed. Head short ; rostrum long and slender, thorax without spines in both sexes, coarsely and closely punctured in wrinkles, sides rather more rounded in female than in male. Rare.

R. Bacchus, Lin.

II. Elytra with more or less regular punctured striae, not wrinkled.

1. Rostrum long and thin ; head more or less sunk in thorax.

A A. Length 2 lines.

Upper-side dark bronze, dull ; under-side darker ; pubescence scanty, whitish-gray. Thorax almost broader than long, closely and rather finely punctured, scarcely dilated at sides ; elytra with deep punctured striae. Not uncommon. *R. cupreus*, Lin.

B B. Length 1—1½ line.

a a. Elytra red, with blackish suture.

Dark bronze ; elytra as above ; base of antennae and legs often red-brown ; pubescence brown. Punctuation extremely close ; thorax with central furrow ; elytra with deep punctured striae. L. 1—1½ l. Common.

R. aequatus, Lin.

b b. Elytra blue, green or bronze.

A a. Upper-side distinctly covered with hairs.

A 1. Elytra with short row of punctures near scutellum, interstices not or scarcely punctured.

a 1. Head and thorax dark bronze or blackish.

Oblong-ovate. Dark bronze, sometimes bluish-black; pubescence scanty, brown. Thorax as broad as long, sides scarcely dilated, finely and very closely punctured; elytra with deep punctured striae, interstices narrow, convex. Rather common. *R. aeneovirens,* Marsh.

b 1. Head and thorax blue.

A 2. Thorax roughly punctured, with central furrow.

Blue, with brown hairs. Thorax roughly punctured, with central furrow; elytra shiny, with strong punctured furrows, interstices narrow, convex, eyes prominent. L. 1¼ l. Not uncommon. *R. pauxillus,* Germ.

B 2. Thorax finely punctured, without central furrow.

Ovate. Greenish-blue, pubescent. Thorax slightly rounded at sides, closely and finely punctured; interstices on elytra impunctate. L. 1¼ l. Common. *R. germanicus,* Herbst.

B 1. Elytra without short row of punctures near scutellum, interstices with a more or less regular row of punctures.

a 1. Thorax longer than broad, widest near base, sides tolerably straight.

Oblong-ovate. Blue or blue-green; antennae, tibiae and tarsi black. Thorax with sides almost straight, narrowed in front, disc not very closely, rather coarsely punctured; elytra with deep punctured striae, interstices flat. L. 1½ l. Rather common. *R. conicus,* Ill.

b 1. Thorax broader than long, widest in middle, sides distinctly rounded.

Ovate. Blue or greenish; slightly pubescent; antennae and legs black. Thorax granulate; closely and finely punctured, with impunctate central line; elytra with deep punctured striae, interstices with a row of minute impressions. L. 1¼—1½ l. Not uncommon. *R. alliariac*, Payk.

B b. Upper-side without or almost without hairs.

Elytra with a short row of punctures near scutellum.

A 1. Tibiae not toothed at apex.

Oblong, rather narrow. Dark greenish-blue. Similar to *R. uncinatus* but with rostrum shorter, angular at base, nearly smooth, thorax roughly punctured, without central furrow. L. 1 l. Not uncommon. *R. nanus*, Payk.

B 1. Tibiae toothed at apex.

Oblong-ovate. Black-blue, shiny, pubescence very fine. Rostrum as long as head and thorax, ridged at base, dilated and wrinkled at apex; head and thorax very finely and indistinctly punctured, latter with a feeble central furrow; elytra with strong punctured striae, interstices scarcely punctured; tibiae toothed at apex. L. 1—1⅛ l. Rather common. *R. uncinatus*, Th.

2. Rostrum short and broad; head long and much narrowed behind eyes.

A A. Length 1½ lines.

Oblong. Blue; closely covered with long hairs. Rostrum scarcely as long as head, straight, feebly ridged; thorax longer than broad, distinctly and rather closely punctured; elytra with shallow punctured striae, interstices with diffusely standing punctures, placed somewhat in rows; tarsi broad, black. Not uncommon.
R. ophthalmicus, Steph.

B B. Length 2½—2¾ lines.

Oblong. Blue; pubescence long, brown. Rostrum longer than head, almost straight, with a central ridge; thorax almost broader than long, sides feebly rounded, very finely and not altogether closely punctured; elytra with punctured striae, interstices flat, with fine punctures, placed somewhat in rows, and somewhat indistinct wrinkles; pygidium covered. Not uncommon.
R. pubescens, Herbst.

b. Head with a circular furrow behind eyes.

I. Elytra blue.

Black, with bronze reflection; elytra blue or blue-green;

pubescence somewhat upright, gray. Head (with eyes) broader than front of thorax, constricted behind, forehead flat, coarsely and diffusely punctured; thorax distinctly longer than broad, finely, rather closely and evenly punctured, sides somewhat rounded; elytra half as long again as together broad, with deep punctured striae, interstices very finely punctured in rows. L. 1½—1¾ l. Not uncommon.   *R. megacephalus*, Germ.

II. Elytra black.

Ovate. Black, shiny; pubescence very fine. Rostrum scarcely longer than head, dilated at apex; head very large, constricted behind, punctured (as also thorax) finely and closely; elytra with punctured striae, interstices with a more or less regular row of punctures. Male with posterior femora much thickened. L. 2 l. Common.

R. betulae, Lin.

ATTELABIDES.

A. Intermediate coxae contiguous or feebly separated by two processes, arising from the meso- and metathorax.

Head more or less prolonged behind eyes, but without constriction or neck; funiculus of antennae with seven joints, first five reversed conical, gradually decreasing; eyes lateral; thorax much shorter than elytra. *Attelabus*, Lin.

B. Intermediate coxae rather strongly separated by a process of metasternum only, reaching the level of their anterior margin.

Head much narrowed and nearly always constricted behind, with a globular neck; joints of funiculus of antennae reversed conical; elytra much broader than thorax; mesosternum without process behind. *Apoderus*, Ol.

Attelabus.

Black; thorax, elytra and usually also base of antennae red; without hairs. Thorax with extremely fine, scattered punctures; elytra with feeble punctured striae, interstices very diffusely punctured. L. 2—2¾ l. Common.

A. curculionoides, Lin.

Apoderus.

Black; hinder part of thorax, elytra and sometimes legs

red ; without hairs. Rostrum, head and thorax with a central furrow ; elytra with coarse and rather irregular punctured striae. L. 3 l. Common. *A. coryli*, Lin.

APIONIDES.

Funiculus of antennae with first joint reversed conical, rest very short ; elytra scarcely broader than thorax in front, dilated behind ; prosternum not emarginate in front ; body broadest behind, much narrowed in front.

Apion, Herbst.

Apion.

A. Rostrum pointed at apex.

a. Rostrum tolerably equal in thickness from base to middle and thence rapidly narrowed toward apex, dilated beneath insertion of antennae.

I. Elytra blue ; antennae entirely black.

Black ; elytra blue. Rostrum dilated beneath insertion of antennae ; head and thorax punctured, latter conical, with a central furrow behind ; elytra ovate, strongly convex, with punctured furrows, interstices flat. L. 1½ l. Common. *A. pomonae*, Fab.

II. Elytra black ; antennae partly red-yellow.

1. Forehead without striae between eyes ; antennae red-yellow until beyond middle.

Black, dull, with gray hairs ; female with base of antennae, male (always much smaller) usually with whole of antennae rust-red. Rostrum dilated beneath insertion of antennae ; thorax conical, with large punctures and a central furrow behind ; elytra short, almost globular-ovate, with punctured furrows, interstices flat. L. 1—1¼ l. Common. *A. craccae*, Lin.

2. Forehead with two striae between eyes ; antennae with only first joint or first two joints red-yellow.

Black, dull, with gray hairs ; first joint of antennae, sometimes also second, yellow. Rostrum much dilated near middle beneath ; forehead striate between eyes ; thorax distinctly bisinuate behind, sides almost straight until middle, then suddenly narrowed in front, strongly and closely punctured, with a central furrow behind ;

elytra ovate, rather convex, with broad punctured furrows, interstices flat, finely wrinkled transversely. L. 1¼ l. Rare.   *A. cerdo,* Gerst.

b. Rostrum gradually narrowed from base to apex, not dilated beneath insertion of antennae.

Black, dull, with fine and very scanty gray pubescence ; base of antennae red-brown. Thorax conical, strongly punctured, with a central furrow behind ; elytra globular-ovate, with punctured furrows, interstices flat. L. 1⅓— 1½ l. Not common.   *A. subulatum,* Kirby.

B. Rostrum not pointed at apex.

a. Antennae inserted at or near base of rostrum.

I. Scutellum small, punctiform.

1. All femora black.

A A. Elytra blue, green or bronze.

a a. Intermediate coxae almost contiguous.

Black, with extremely fine gray pubescence ; elytra black-blue. Rostrum dilated into a little tooth on each side, above base of antennae ; forehead with fine longitudinal wrinkles ; thorax almost longer than broad, with large, round, deep punctures, sides straight, somewhat narrowed in front ; elytra with deep, distinctly punctured striae. L. 1¼ l. Common.   *A. carduorum,* Kirby.

b b. Intermediate coxae placed apart.

Black, without pubescence ; elytra bronze, green or blue-green. Rostrum long, scarcely dilated above base of antennae ; forehead wrinkled ; thorax cylindrical, as long as broad, with large, round, deep punctures, with a short central furrow behind ; elytra oval, with deep indistinctly punctured striae. L. 1¼—1⅓ l. Common.   *A. onopordi,* Kirby.

B B. Elytra black or leaden-black, not metallic (at least in male)

a a. Forehead with two furrows, united behind.

A a. Striae on elytra distinctly punctured.

Very similar to *A. confluens,* but with pubescence still feebler, elytra ovate, with deep, distinctly punctured striae. L. ¾ l. Moderately common.   *A. stolidum,* Germ.

B b. Striae on elytra rather indistinctly punctured.

Narrow. Black, with very fine, gray pubescence. Forehead with two short, deep furrows (united behind) between eyes; thorax almost cylindrical, as long as broad, very finely and scantily punctured, with a little depression before scutellum ; elytra oblong-ovate, with fine, rather indistinctly punctured striae. L. 1 l. Rather common.

A. confluens, Kirby.

b b. Forehead with four or five striae between eyes.

Black, not very convex, bare ; elytra of male black, of female violet. Head with four or five striae between eyes ; thorax cylindrical, in middle smooth, sides punctured ; elytra ovate, with fine punctured striae. L. 1¼ l. Rare.

A. laevigatum, Kirby.

c c. Forehead without furrows or striae.

A a. Thorax transverse.

A 1. Antennae entirely black.

a 1. Shoulders of elytra obtusely prominent; length 1 line.

Black, with scanty gray pubescence. Head and thorax strongly punctured, latter distinctly broader than long, sides rounded, narrowed in front ; elytra much broader than thorax, shoulders obtusely prominent, with deep punctured furrows, interstices flat. Moderately common. *A. vicinum*, Kirby.

b 1. Shoulders of elytra bluntly rounded ; length ⅔ line.

Very similar to *A. vicinum* but with closer and longer pubescence, rostrum longer and more curved, sides of thorax more rounded, elytra rather narrower. Moderately common.

A. atomarium, Kirby.

B 1. Antennae more or less yellow-brown.

a 1. Rostrum shiny and very finely punctured.

Similar to *A. flavimanum* but with thorax rather shorter and not so closely punctured, interstices on elytra not so dull. Antennae of female yellow-brown at base, of male entirely yellow-brown, except club ; legs of female entirely black, much stouter than in *A. flavimanum*, with wider tarsi, of male with femora (especially anterior pair) very robust and all tibiae marked with yellow-brown near base. L. ⅘ l. Rare.

A. annulipes, Wenck.

b 1. Rostrum closely punctured and pubescent.

Black, with gray pubescence ; base of antennae and tibiae yellow-brown. Rostrum not much longer than head, thick, not much curved, closely punctured and pubescent ; thorax rather broader than long, closely and distinctly punctured, base bisinuate, with angles prominent ; elytra ovate, with very diffusely punctured furrows. L. $\frac{4}{5}$ l. Not uncommon.   *A. flavimanum,* Gyll.

B b. Thorax nearly globular.

Black, dull; with feeble pubescence ; elytra with slightly greenish reflection. Rostrum moderately long, somewhat thickened at base of antennae ; thorax as broad as long, closely and finely punctured, with a scarcely visible central furrow behind, sides rounded ; elytra oblong, with deep punctured striae, interstices flat, transversely wrinkled. L. $1\frac{1}{4}$ l. Rather common.   *A. Hookeri,* Kirby.

2. At least anterior femora brown-red or red-yellow.

A A. Rostrum dilated into a little tooth above insertion of antennae ; intermediate coxae placed near each other.

a a. Pubescence on elytra evenly distributed.

Black, with rough gray hairs ; anterior legs brown-red, base of femora black, tibiae of hinder pairs red ; sometimes with intermediate pair or all legs red. Rostrum very long, thin, deep black ; thorax almost globular, rather broader behind, narrowly bordered in front, with a central furrow (hidden by pubescence) before scutellum. L. $1\frac{1}{2}$ l. Common.   *A. ulicis,* Forst.

b b. Elytra with markings of closer pubescence.

A a. Antennae at least partly red-yellow.

A 1. White streak on elytra oblique.

Very similar to *A. genistae,* but with rostrum longer, thorax almost longer than broad, elytra usually brown, their streak of white hairs always abbreviated and reaching obliquely from shoulder to middle of elytron, legs generally yellowish-red. L. 1—$1\frac{1}{2}$ l. Common. *A. fuscirostre,* Fab.

B 1. White streak on elytra longitudinal.

Black, upper-side covered with yellowish-gray recumbent hairs, sides of thorax, a straight broad longitudinal streak on middle of each elytron and under-side more closely covered with gray-white hairs; antennae (or their base only) and legs reddish-yellow. Thorax distinctly broader than long, narrowed in front; elytra rather broader at base than thorax, scarcely dilated behind, striae rather indistinct on account of pubescence. L. $\frac{4}{5}$ l. Rather common.

A. genistae, Kirby.

B b. Antennae entirely black.

Black, with scanty gray pubescence, each elytron with an abbreviated streak of more closely placed white hairs at base; legs pale red-yellow. Rostrum thin, curved, thorax short, cylindrical; elytra oblong-ovate. L. 1—1¼ l. Not common. *A. semivittatum,* Gyll.

B B. Rostrum not dilated above insertion of antennae; intermediate coxae placed apart.

a a. Elytra black, evenly pubescent throughout.

Black, not very shiny, with scanty gray pubescence; base of antennae and legs reddish-yellow-brown, base of femora, trochanters, knees and tarsi blackish. Rostrum cylindrical, curved; forehead with a central furrow; thorax finely and very closely punctured, with a short central furrow behind; elytra oblong-ovate, with deep punctured striae. L. 1 l. Rather common. *A. pallipes,* Kirby.

b b. Elytra brown, with two bare oblique bands.

Pitch-black, with close, whitish-gray pubescence; antennae and legs reddish-yellow; elytra brown, with two more or less regular, hairless, oblique bands. L. 1 l. Common. *A. vernale,* Fab.

II. Scutellum linear.

1. Forehead with a deep longitudinal impression.

Black, shiny, without pubescence; elytra bronze, green, or blue-green. Forehead with a deep longitudinal depression between eyes; thorax conical, somewhat longer than broad, finely punctured, with a short central furrow or a depression behind; elytra oval, with very fine simple striae, interstices flat. L. 1⅓—1½ l. Common. *A. aeneum,* Fab.

2. Forehead without impression.

Black, shiny, with very fine and scanty pubescence ; elytra bluish-green or black-green. Forehead wrinkled, without longitudinal depression, vertex flattened transversely ; thorax very finely punctured ; elytra with rather deep, indistinctly punctured striae ; interstices slightly convex, with very fine, scarcely visible punctures. L. 1¼ —1⅓ l. Common.   *A. radiolus*, Kirby.

b. Antennae inserted at about basal third-part of rostrum.

I. Legs black.

1. Elytra without pubescence.

Green or greenish-blue, shiny ; rostrum black, shiny. Forehead wrinkled ; eyes not very prominent ; thorax as long as broad, with large, deep punctures and a short, deep central furrow ; elytra ovate, with fine punctured striae. L. 1 l. Rare.   *A. astragali*, Payk.

2. Elytra with more or less distinct pubescence.

A A. Elytra very strongly inflated behind middle, not narrowed behind, pear-shaped.

a a. Thorax with central furrow nearly from base to apex.

Black, dull, with' very feeble pubescence. Head finely punctured, nearly smooth ; rostrum rather long, curved ; thorax as long as broad, cylindrical, closely, deeply and rather coarsely punctured, with a central furrow nearly throughout ; elytra with deep, broad punctured furrows, interstices convex. L. 1 l. Common. *A. striatum*, Kirby.

b b. Thorax with an impression before scutellum.

Black, with very fine gray pubescence. Head wrinkled between eyes ; rostrum feebly punctured, shiny ; thorax globular, very closely and finely punctured, with an impression before scutellum ; elytra with punctured furrows, interstices flatly arched. L. 1—1½ l. Rather common.   *A. immune*, Kirby.

B B. Elytra ovate or oblong.

a a. Forehead with depression.

A a. Head deeply impressed between eyes ;
thorax with a short central furrow.

Elongate. Black, with whitish pubescence. Rostrum
of male rather longer than head, thick ; that of female
longer and thinner ; forehead feebly striate; thorax tolerably cylindrical, scarcely as long as broad, with a short
central furrow, base indistinctly emarginate ; elytra long
oval, with punctured striae. L. 1 l. Moderately common.
A. pubescens, Kirby.

B b. Head faintly impressed between eyes ;
thorax without central furrow.

Black ; antennae pitchy, club black ; glabrous, slightly
shiny. Forehead feebly striate; thorax slightly elongate,
punctured ; elytra sometimes with a bluish tinge. L. 1¼ l.
Moderately common. *A. Curtisi,* Curt.

b b. Forehead without depression.

A a. Elytra black-bronze ; thorax roughly
punctured.

Black, slightly shiny, with thin gray pubescence; elytra
with metallic reflection. Thorax narrowed in front, closely
punctured, with a short, feeble central furrow behind ;
elytra oblong, narrow, with deep punctured striae and
studded with isolated, long, white hairs. L. 1¼ l. Not
common. *A. simile,* Kirby.

B b. Elytra leaden-black ; thorax feebly
punctured.

Black, with fine gray pubescence. Head narrow; thorax
as long as broad, feebly punctured, with an impression
before scutellum ; elytra oblong, tolerably straight at sides,
only slightly dilated behind middle, interstices flat. L. ⅘ l.
Common. *A. seniculum,* Kirby.

II. Legs reddish-yellow.

Black ; elytra green or blue, shiny ; antennae and legs
reddish-yellow ; upper-side scantily, under-side closely
covered with whitish pubescence. Thorax short, broader
behind ; elytra with indistinctly punctured striae. Rostrum
of male shorter than that of female and half of it (as also
apex of abdomen) red-yellow. L. 1 l. Common.
A. rufirostre, Fab.

c. Antennae inserted in or near middle of rostrum.

I. Reflexed margin of elytra not sinuate behind posterior coxae; second interstice recurved hooklike at apex.

 1. At least anterior femora red or red-yellow.

  A A. Elytra pubescent.

Black; antennae yellow, club black in female; femora and tibiae yellow, tarsi black; upper-side with scanty, under-side with closer grayish-white pubescence. Rostrum as long as thorax, latter as long as broad, narrowed in front, strongly punctured, with a short central furrow behind; elytra oval, with punctured furrows, interstices flat. L. 11. Common.   *A. viciae,* Payk.

  B B. Elytra not pubescent.

   a a. Hinder pairs of tibiae partly or wholly black.

    A a. Third and fourth joints of antennae of male very broad.

Black, shiny, bare; male with antennae yellowish-brown, black toward apex, first joint clublike, second small, third and fourth very broad, next four small, femora and a ring on tibiae yellowish-red; female with antennae simple, black, femora yellowish-red, trochanters and tibiae black. Thorax conical, closely and moderately strongly punctured, with an indistinct central furrow behind; elytra oval, with punctured furrows. L. 1 l. Moderately common.   *A. difforme,* Ahr.

    B b. Third and fourth joints of antennae of male not broad.

     A 1. Sides of thorax distinctly rounded.

Black, shiny, almost bare; femora and base of tibiae red; first joint of antennae yellow in male. First joint of antennae elongate (more so in female than in male), thickened clublike at apex; rostrum curved, shorter and thicker in male than in female; thorax narrow, closely punctured, rounded at sides; elytra oval, with fine punctured striae, interstices flat, impunctate. L. 1⅓ l. Not uncommon.   *A. dissimile,* Germ.

     B 1. Thorax cylindrical.

      a 1. Third joint of antennae elongate conical.

A 2. Thorax feebly punctured.

a 2. Base of antennae and of hinder pairs of tibiae red-yellow.

Black, shiny, bare; base of antennae and legs red-yellow, latter with apex of posterior tibiae and tarsi black. Forehead with a short furrow; thorax oblong, finely, rather closely punctured, with a central furrow behind; elytra oblong oval, with moderately strong punctured furrows. L. 1¼ l. Rather common. *A. laevicolle,* Kirby.

b 2. Antennae and hinder pairs of tibiae entirely black or black-brown.

Oblong. Black, bare; legs red, posterior tibiae and tarsi black-brown. Rostrum somewhat longer than thorax, curved; thorax narrow, nearly cylindrical, indistinctly and rather diffusely punctured, with a small, feeble depression at base; elytra with tolerably distinct punctured striae, interstices flat, finely wrinkled transversely. L. ⅘ l. Rare. *A. Schoenherri,* Boh.

B 2. Thorax strongly punctured.

a 2. Rostrum with a fine ridge at base.

Black, bare; first three joints of antennae, femora and anterior tibiae red-yellow. Forehead punctured and striate between eyes, the fine central ridge extending nearly to insertion of antennae; rostrum long in female; thorax oblong, strongly punctured in wrinkles, with central furrow at base; elytra oblong-ovate, with punctured furrows; fourth joint of antennae oblong; anterior tibiae straight. L. 1½ l. Rather common. *A. ononidis,* Germ.

b 2. Base of rostrum without ridge.

A 3. Hinder pairs of tibiae yellow at base.

Black, bare; base of antennae yellow-brown, darker in female; legs yellow, knees and apical half of tibiae blackish. Rostrum long, curved, diffusely punctured; forehead punctured between eyes, especially behind, scarcely striate in front; thorax oblong, strongly punctured, with a shallow central furrow behind; elytra oblong oval, convex, with punctured striae; third joint of antennae nearly double as long as fourth; anterior tibiae curved, very slightly in female, more strongly in male. L. 1 l. Moderately common.   *A. varipes*, Germ.

> **B 3.** Hinder pairs of˙ tibiae entirely black.

Black, bare; base of antennae, femora and anterior tibiae yellow-brown. Forehead punctured in wrinkles; rostrum long, slightly curved; thorax longer than broad, almost cylindrical, not very closely, rather strongly punctured, with central furrow at base; elytra with punctured striae, interstices slightly convex; anterior tibiae straight. L. 1—1¼ l. Common.   *A. fagi*, Lin.

> **b 1.** Third joint of antennae short conical.

> **A 2.** Trochanters reddish.

Narrow. Black, bare; base of antennae brown; femora red. Thorax cylindrical, narrow, closely and finely punctured, with feeble central furrow behind; elytra oblong ovate, with rather indistinctly punctured striae, interstices flat. L. 1 l. Common.   *A. assimile*, Kirby.

> **B 2.** Trochanters black or blackish.

Black, scarcely pubescent, femora red. Head short and broad; rostrum very little curved; thorax cylindrical, not very closely punctured; elytra ovate, convex, with deep punctured striae, interstices slightly convex. L. ¾—1 l. Common.   *A. trifolii*, Lin.

> **b b.** All tibiae red or yellow.

> **A a.** Male with rostrum entirely black; female with anterior coxae black.

Black, with extremely fine pubescence; base of antennae reddish-yellow; femora and tibiae red. Forehead with deep longitudinal wrinkles; thorax as long as broad, slightly narrowed in front, strongly and rather closely punctured, with a central furrow behind; elytra oval, convex, with deep punctured striae. L. ¾—1 l. Common.
                                        A. flavipes, Fab.

> **B b.** Male with apex of rostrum yellow; female with anterior coxae red.

Black, rather shiny, pubescence extremely fine; antennae

(except club) and legs (except tarsi) yellow. Forehead very finely wrinkled; thorax almost broader than long, very finely punctured; elytra oval, very convex, with punctured furrows. Rostrum of male shorter than that of female and with anterior half yellow. L. ¾ l. Common.
A. nigritarse, Kirby.

2. Legs entirely black.

A A. Thorax feebly punctured.

a a. Shoulders of elytra not prominent.

Black, shiny, bare. Rostrum distinctly punctured; forehead striate; thorax oblong, nearly cylindrical, diffusely punctured, with a short broad central furrow behind; elytra oval, with punctured furrows, interstices convex; antennae short. L. 1¼ l. Moderately common.
A. ebeninum, Kirby.

b b. Shoulders of elytra prominent.

A a. Elytra black.

Deep black, rather shiny, bare. Rostrum moderately long, shiny, antennae inserted nearer its base in male than in female; forehead punctured, punctures confluent; thorax oblong, cylindrical, diffusely and feebly punctured, with an impression before scutellum; elytra oblong oval, narrow, with fine punctured striae, interstices flat. L. 1 l. Rather common.
A. tenue, Kirby.

B b. Elytra blue or green.

A 1. Antennae inserted in middle of rostrum; first tarsal joint distinctly longer than second.

Black, bare; elytra dark blue. Rostrum thickened at base, shiny in front; forehead furrowed; thorax as long as broad, almost cylindrical, with fine indistinct punctures and a feeble central furrow behind; elytra oval, with punctured furrows, interstices slightly convex. L. 1¼ l. Rather common.
A. punctigerum, Payk.

B 1. Antennae inserted a little behind middle of rostrum; first tarsal joint only slightly longer than second.

Black, shiny; elytra greenish-blue or green. Forehead wrinkled; thorax much broader than long, usually with metallic reflection, extremely feebly punctured, with a small, puncturelike impression before scutellum; elytra

N 2

oval, with punctured striae, sutural stria deeper than the
rest. L. 1—1¼ l. Common.   *A. virens*, Herbst.

> **B B.** Thorax distinctly punctured.
>
> > *a a.* Elytra black or leaden-black.
> >
> > > *A a.* Scutellum small, punctiform.
> > >
> > > > *A 1.* Rostrum rather long.
> > > >
> > > > > *a 1.* First tarsal joint distinctly longer
> > > > > than second ; thorax with central
> > > > > furrow reaching at least to middle.
> > > > >
> > > > > *A 2.* Eyes not strongly prominent ;
> > > > > antennae entirely black, or dusky
> > > > > yellow at base only.
> > > > >
> > > > > > *a 2.* Central furrow on thorax
> > > > > > reaching to middle.
> > > > > >
> > > > > > > *A 3.* Vertex of head short;
> > > > > > > rostrum of male somewhat
> > > > > > > dilated in middle and slightly
> > > > > > > thickened beneath.

Black, almost bare ; base of antennae yellow-red. Rostrum long, curved ; head flat, nearly square ; thorax nearly
cylindrical, narrow, closely and moderately strongly punctured, with a central furrow behind ; elytra ovate, with
punctured furrows, interstices flat, finely punctured. L.
1¼ l. Common.   *A. platalea*, Germ.

> > > > > > > *B 3.* Vertex of head long ; ros-
> > > > > > > trum of male somewhat thick-
> > > > > > > ened beneath before insertion
> > > > > > > of antennae.

Black, finely pubescent. Head narrow ; rostrum very
long, distinctly punctured ; eyes not convex ; forehead
finely striate ; thorax almost cylindrical, strongly punctured, with a short central furrow behind ; elytra oblong,
with deep punctured striae, interstices flat. L. 1 l. Moderately common.   *A. Gyllenhali*, Kirby.

> > > > > *b 2.* Central furrow on thorax
> > > > > reaching from base to apex.

Similar to *A. ervi*, but with antennae entirely black,
pubescence rather closer, rostrum punctured in wrinkles,
central furrow of thorax somewhat longer. L. 1¼ l. Common.   *A. ononis*, Kirby.

B 2. Eyes strongly prominent; antennae with at least basal half red-
yellow.

Black, with scanty pubescence; antennae with base in
female, in male entirely yellow. Rostrum elongate, curved;
forehead striate; thorax scarcely as long as broad, strongly
punctured, with a central furrow behind; elytra oval, with
punctured striae, interstices flat. L. ⅘ l. Common.

A. ervi, Kirby.

b 1. First tarsal joint not or only
slightly longer than second; thorax
with a basal depression or central
furrow not reaching middle.

A 2. Elytra not pubescent.

Deep black, rather shiny, almost bare. Rostrum and
legs elongate; thorax cylindrical, as broad as long, with
large, deep punctures and a short central furrow behind;
elytra nearly globular, with punctured furrows, interstices
broader than furrows, flat. L. 1¼ l. Scarce.

A. filirostre, Kirby.

B 2. Elytra pubescent.

Black, finely pubescent. Rostrum long, thin, shiny,
scarcely punctured; forehead wrinkled; thorax cylindrical,
strongly punctured, with a fine central furrow behind;
elytra oblong oval, with broad punctured furrows. L. 1—
1¼ l. Common. *A. loti*, Kirby.

A 1. Rostrum short.

Black, rather dull, almost bare. Rostrum rather longer
than head, smooth; forehead wrinkled; thorax as long as
broad, closely and strongly punctured, with a feeble depression before scutellum; elytra oblong oval, with broad and
deep punctured striae, interstices scarcely broader than
striae, convex. L. ⅘ l. Moderately common.

A. minimum, Herbst.

B b. Scutellum elongate.

Black, rather narrow. Distinguished from *A. meliloti* by
having the rostrum rather longer, body more hairy, eyes
less prominent, forehead without depression and elytra
black, with wider furrows. L. 1⅔ l. Not very common.

A. scutellare, Kirby.

b b. Elytra blue or green.
A a. Elytra not pubescent.
A 1. Sides of thorax straight; first tarsal
joint half as long again as second.
a 1. Eyes prominent.

Black, bare ; elytra dark-blue.   Rostrum and head closely
and deeply punctured ; eyes prominent ; thorax nearly
cylindrical, deeply and distinctly punctured, with a short
central furrow behind ; elytra short oval, convex, with
punctured furrows, interstices rather convex.   L. 1¼ l.
Common.   *A. pisi*, Fab.

b 1. Eyes not prominent.

Black, bare ; elytra black-blue.   Rostrum rather long
and thin, punctured ; forehead wrinkled ; eyes not prominent ; thorax cylindrical, as long as broad, rather diffusely
punctured, usually with a shallow central furrow ; elytra
short oval, convex, with punctured furrows, interstices flat.
L. 1 l.   Common.   *A. aethiops*, Herbst.

B 1. Sides of thorax distinctly rounded ;
first tarsal joint equal to second.

Black, bare ; elytra of female black-blue.   Rostrum
cylindrical, curved, in female as long as body, in male
rather shorter ; eyes not prominent ; forehead wrinkled ;
thorax rounded at sides, strongly but not very closely
punctured, with a deep furrow before scutellum ; elytra
with very fine, indistinctly punctured striae, interstices flat
and more than three times as broad as striae.   L. 1½ l.
Not common.   *A. sorbi*, Herbst.

B b. Elytra pubescent.

A 1. Elytra elongate ; length 1¾ lines.

Black, finely pubescent ; elytra greenish-blue.   Head
elongate ; forehead striate , rostrum long, much curved ;
thorax oblong, cylindrical, closely and distinctly punctured,
with a central furrow behind ; elytra elongate oval, very
convex behind, with punctured furrows, interstices flat.
Not uncommon.   *A. meliloti*, Kirby.

B 1. Elytra not elongate ; length 1—1½
lines.

a 1. Forehead with two or three impressed striae.

A 2. Antennae entirely black.

Black, dull, pubescent; elytra blue. Rostrum rather thick; forehead with three striae; thorax rounded at sides, narrowed in front, closely and distinctly punctured, with a central furrow; elytra oval, with punctured furrows, interstices flat. L. 1½ l. Not uncommon. *A. Spencei*, Kirby.

B 2. Antennae red-yellow at base.

Black, pubescent; elytra black-blue; base of antennae brown-red. Apical half of rostrum shiny; forehead striate; eyes not very prominent; thorax somewhat conical, narrowed in front, as long as broad, closely and distinctly punctured, with a central furrow; elytra oblong oval, very convex behind, with punctured furrows. L. 1—1¼ l. Common. *A. vorax*, Herbst.

b 1. Forehead punctured.

A 2. Elytra oval; length 1 line.

Leaden-black, almost bare; elytra black-blue. Rostrum almost longer than head and thorax, curved, punctured; thorax nearly cylindrical, scarcely longer than broad, base and apex truncate and bordered, closely and distinctly punctured, with a short central furrow behind; elytra oval, with close punctured furrows, interstices flat. Not uncommon. *A. livescerum*, Gyll.

B 2. Elytra short oval; length 1⅓ lines.

Black, pubescent; elytra dark blue. Apex of rostrum shiny; forehead finely punctured in rows, with a narrow impunctate central line (often indistinct); eyes somewhat prominent; thorax nearly cylindrical, closely and distinctly punctured, with a short central furrow; elytra short oval, very convex behind, with punctured furrows, interstices flat. L. 1⅓ l. Not uncommon. *A. Waltoni*, Steph.

II. Reflexed margin of elytra sinuate behind posterior coxae, second interstice recurved hooklike at apex.

1. Body entirely red or yellowish-red.

A A. Elytra dilated behind.

a a. Thorax with a central furrow behind.

Blood-red; eyes black. Vertex of head and temples deeply and closely punctured, former long. Rostrum rather thick, curved; thorax broader than long, rounded at sides, bordered in front, closely punctured, with a fine central furrow behind; elytra oblong oval, with deep punctured striae.   L. 1¾ l.   Common.   *A. miniatum*, Germ.

b b. Thorax with a depression at base only.

A a. Thorax somewhat dilated at sides.

Yellow-red, with feeble pubescence; eyes black.   Head rather long, coarsely punctured in wrinkles; eyes very prominent; rostrum very thick, short, curved, punctured, shiny; thorax almost cylindrical, sides somewhat dilated, narrowed and bordered in front, coarsely punctured, without central furrow; elytra very convex, with deep punctured furrows, interstices raised, almost as broad as furrows.   L. 1½—1¾ l.   Not common.   *A. cruentatum*, Walt.

B b. Thorax not dilated at sides.

Yellowish-red, dull; legs paler; eyes black.   Vertex of head and temples deeply punctured.   Rostrum short, curved, impunctate; thorax strongly punctured, without central furrow; elytra with notched striae.   L. 1¼—1½ l.   Common.   *A. frumentarium*, Lin.

B B. Sides of elytra nearly parallel.

a a. Rostrum curved, shorter.

Narrow.   Light blood-red, dull, pubescent; legs paler; eyes black.   Vertex of head rather short, not very deeply punctured; temples not punctured; rostrum rather long, thin, curved; thorax slightly transverse, rather finely punctured, with only a basal depression, sides nearly straight; elytra with notched striae.   L. 1¼ l.   Rather common.

A. rubens, Steph.

b b. Rostrum straight, longer.

Yellowish-red, dull, slightly pubescent; legs paler, eyes black; vertex of head not punctured at sides; rostrum straight, in male somewhat, in female much longer than thorax; thorax not transverse, closely punctured, with only a small basal depression, sides not dilated; elytra with notched striae.   L. 1¼—1½ l.   Rare.

A. sanguineum, De G.

2. Body wholly or partly black or bronze.

A A. Elytra yellow-brown, with a basal spot, suture and side margin darker.

Black, with close whitish-gray rough pubescence; elytra yellow-brown, a triangular common spot at base, suture and side margin darker, antennae and legs reddish-yellow. Rostrum short, thick, only slightly curved (rather longer, thinner and more curved in female than in male); thorax nearly conical, closely punctured; elytra with punctured striae. L. 1 l. Common. *A. malvae*, Fab.

B B. Elytra black or blue.

a a. Thorax copper.

Bronze; upper-side purplish-copper, with fine recumbent pubescence; eyes and antennae black. Head smooth behind, rostrum rather thick, forehead and thorax punctured; elytra globose ovate, with punctured striae. L. 1 —2 l. Moderately common. *A. limonii*, Kirby.

b b. Thorax black.

A a. Thorax diffusely punctured, with central line impunctate.

Black, scantily pubescent. Head elongate, punctured in wrinkles; forehead with two or three furrows; rostrum short, marked with a fine stria; thorax nearly cylindrical, as long as broad, disc diffusely, sides not closely punctured, with an impression before scutellum; elytra oblong oval, with narrow, feebly punctured furrows, interstices flat. L. 1 l. Scarce. *A. sedi*, Germ.

B b. Thorax closely punctured.

A 1. Elytra blue or greenish-blue.

a 1. Sides of thorax not or scarcely rounded.

A 2. Funiculus of antennae thick, second joint almost transverse.

a 2. Thorax with only a depression before scutellum.

Black, nearly bare; elytra blue. Rostrum short, curved, punctured; forehead wrinkled; thorax as long as broad, cylindrical, shallowly and not closely punctured, with a rather deep oblong depression before scutellum; elytra elongate oval, with punctured furrows. L. 1¼ l. Common. *A. violaceum*, Kirby.

b 2. Thorax with a feeble central furrow behind.

Similar to *A. violaceum*, but with rostrum rather shorter and thicker at base, antennae inserted somewhat nearer its base, thorax with a feeble central furrow behind. L. 1¼ l. Moderately common.   *A. hydrolapathi*, Kirby.

> *B 2.* Funiculus of antennae thin, second joint nearly oval.

Black, rather shiny, nearly bare ; elytra blue. Rostrum of male short ; thorax narrow, not very closely punctured ; elytra oval, with punctured furrows. L. 1¼ l. Moderately common.   *A. marchicum*, Herbst.

> *b 1.* Sides of thorax distinctly rounded.

Black, rather shiny, nearly bare ; elytra greenish-blue. Head large, punctured ; rostrum short, very thick ; thorax almost globular, very closely punctured, with a short central furrow or a depression before scutellum ; elytra oval, with punctured striae, interstices flat. L. 1—1½ l. Moderately common.   *A. affine*, Kirby.

> *B 1.* Elytra black.

Black, with gray pubescence. Rostrum very short ; forehead finely wrinkled ; thorax cylindrical, as broad as long, distinctly punctured, with an impression behind ; elytra oblong oval, with punctured furrows, interstices very slightly convex. L. 1 l. Common.   *A. humile*, Germ.

ERIRHINIDES.

A. Eyes small, rounded.

Antennae short, scape not reaching over eyes, funiculus with seven joints; thorax slightly bisinuate at base; second abdominal segment scarcely longer than third, separated from first by a straight suture ; tarsi short, rather narrow, fourth joint reaching moderately beyond anterior margin of third, claws very small, soldered at base.

> *Brachonyx*, Schoenh.

B. Eyes moderately large, transverse.

> *a.* Fourth tarsal joint not or scarcely reaching beyond front of third.

> *I.* Funiculus of antennae with seven joints.

> *1.* Fourth tarsal joint absent.

Rostrum scarcely longer than head, scrobes oblique ;

antennae rather short, first two joints of funiculus elongate; scutellum elongate triangular; second abdominal segment longer than third and fourth together, separated from first only by a very fine curved suture; tarsi short, third joint bilobed, much broader than first and second.

Anoplus, Schoenh.

2. Fourth tarsal joint present.

Rostrum elongate, scrobes oblique; antennae moderately long, first joint of funiculus elongate, rest very short; scutellum scarcely visible; second abdominal segment as long as third and fourth together, separated from first by a straight suture; prosternum strongly emarginate in front; tarsi short, spongy beneath, third joint bilobed, broader than first and second. *Smicronyx*, Schoenh.

II. Funiculus of antennae with six joints.

Antennae moderately long, first two joints of funiculus somewhat elongate; second abdominal segment as long as third and fourth together, separated from first by a curved suture; tarsi short, spongy beneath, third joint bilobed, broader than first and second. *Tanysphyrus*, Schoenh.

b. Fourth tarsal joint reaching much beyond front of third.

I. Tarsi very slender, not spongy beneath, third joint not or scarcely broader than first and second, not bilobed.

1. Funiculus of antennae with second joint longer than first.

Antennae moderately long, funiculus with six or seven joints; prosternum more or less hollowed out; second abdominal segment distinctly longer than third and fourth together, separated from first by a straight suture.

Bagous, Schoenh.

2. Funiculus of antennae with first joint longer than second.

Similar to *Bagous*, but with club of antennae less compact, prosternum not hollowed out and scutellum larger.

Hydronomus, Schoenh.

II. Tarsi more or less broad, spongy beneath, third joint broader than first and second, bilobed.

1. Scrobes commencing at some distance from mouth.

A A. Funiculus of antennae with five joints.

Antennae short, moderately robust; eyes rather small; elytra elongate, parallel-sided, not broader than thorax; intermediate segments of abdomen equal, second separated from first by a straight suture; tibiae straight, tarsi rather short, claws very small.   *Mecinus*, Germ.

B B. Funiculus of antennae with seven joints.

a a. Anterior tibiae more or less curved.

Antennae more or less long, slender; elytra oval or oblong oval, a little broader than thorax; second abdominal segment at least as long as third and fourth together, separated from first by a nearly straight suture; tarsi rather long and narrow, claws moderately large.

Erirhinus, Schoenh.

b b. Anterior tibiae straight.

Antennae rather long, slender; eyes rather large; elytra somewhat short, nearly parallel-sided, distinctly broader than thorax; second abdominal segment longer than third and fourth together, separated from first by a curved suture, tarsi moderately broad, claws rather large.

Grypidius, Schoenh.

2. Scrobes commencing close to mouth.

Antennae placed far forward, moderately long, rather robust, funiculus with seven joints, first two of them elongate, third also conical but shorter, rest short; eyes rather large; elytra oblong oval, a little broader than thorax; second abdominal segment much longer than third and fourth together, separated from first by a fine straight suture; tibiae straight, rounded at apex, tarsal claws long.

Procas, Steph.

Brachonyx.

Narrow. Reddish-yellow-brown; rostrum, breast, abdomen and often also head and thorax blackish; pubescence yellowish-gray. Thorax closely punctured; elytra with deep punctured striae. L. 1—1¼ l. Rare.

B. indigena, Herbst.

Anoplus.

A. Interstices on elytra punctured in wrinkles.

Ovate. Black, shiny, scantily covered with short whitish hairs, scutellum and breast with closer grayish pubescence ; scape of antennae yellow-brown. Thorax closely and strongly punctured, with a fine, somewhat raised central line ; elytra with deep punctured striae, interstices punctured in wrinkles. L. 1 l. Common.

A. plantaris, Naez.

B. Interstices on elytra with tolerably regular rows of punctures.

Similar to A. plantaris but larger, thorax with coarser punctures, partly confluent, striae on elytra less deep, interstices broader, flat, each with a tolerably regular row of fine punctures bearing short, white bristles, inclined backward. Rare.

A. roboris, Suffr.

Smicronyx.

A. Length 1 line.

a. Thorax shallowly punctured.

Oblong-ovate. Black ; base of antennae red-yellow ; covered with whitish scales, on upper-side scantily, in spots, on under-side closely. Thorax convex, shiny, shallowly punctured, scarcely constricted in front ; elytra with striae distinct but obsoletely and diffusely punctured, interstices flat, punctured. L. 1 l. Scarce.

S. jungermanniae, Reich.

Of S. cicur, Reich I fail to find any description.

b. Thorax distinctly punctured.

Oblong-ovate, convex. Black ; under-side covered with white scales, placed more closely at sides. Rostrum distinctly punctured, not shiny ; thorax distinctly and very closely punctured, broadly constricted before apex, with an indistinct central ridge ; elytra with fine, diffusely punctured striae, shoulders raised, nearly angular. L. 1 l. Rare.

S. Reichi, Gyll.

B. Length ¾ line.

Deep black, shiny, upper-side sparingly, under-side and legs closely covered with minute white scalelike hairs, elytra a little variegated toward apex. Thorax with vario-

lose punctures ; elytra distinctly striated. L. ¾ l. Rare.
S. pygmaeus, Curt.

Tanysphyrus.

Pitch-black ; antennae and legs usually brown ; sides of
thorax and some more or less distinct spots on elytra
covered with gray scales. Elytra with deep punctured
striae, interstices narrow, strongly raised. L. ⅔ l. Common.   *T. lemnae*, Fab.

Bagous.

A. Funiculus of antennae with seven joints.

a. Antennae inserted a little before middle of rostrum ;
third tarsal joint not broader than second.

I. Sides of thorax nearly straight, not rounded after
apical constriction.

1. Elytra with tubercles before apex ; tibiae not
dilated on inner-side above middle.

A A. Length 2½ lines.

a a. Each elytron with two tubercles before
apex.

Oblong. Black ; antennae, tibiae and tarsi rust-red ;
closely covered with brown scales. Forehead with a depression ; rostrum short, thick, moderately curved ; thorax
not transverse, constricted toward apex, nearly straight at
sides, extremely closely punctured, with an indistinct central
furrow ; elytra with fine punctured striae, alternate interstices raised, second and fourth with a tubercle behind.
Rare.   *B. binodulus*, Herbst.

b b. Each elytron with only one tubercle
before apex.

Black ; closely covered with gray scales. Thorax
broader than long, sides moderately dilated, very closely
punctured, with a short, indistinct central furrow behind ;
elytra with punctured striae, interstices unequally granulate, the alternate ones raised, fourth with a tubercle
behind. Rare.   *B. nodulosus*, Gyll.

B B. Length 1½—1¾ lines.

a a. Thorax with an indistinct central ridge
in front.

Black ; tibiae and tarsi red-brown ; covered with gray scales, two broad dorsal spots on thorax, middle of disc and apex of elytra bare, third interstice with a round spot of white scales a little behind middle. Thorax feebly constricted toward apex, closely punctured, with an indistinct central ridge in front ; elytra with fine punctured striae, alternate interstices somewhat raised ; third tarsal joint slightly dilated. L. 1¾ l. Scarce. *B. subcarinatus*, Gyll.

b b. Thorax with a central furrow.

Oblong-ovate. Black ; base of antennae, tibiae and tarsi brown-red ; checkered with gray scales, placed more closely and lighter coloured on head and sides of thorax and elytra. Thorax broader than long, broadly constricted toward apex, closely and distinctly granulate, with a central furrow ; elytra with fine punctured striae, interstices somewhat convex ; tarsi with second joint nearly transverse, as broad as third. L. 1½ l. Moderately common.

B. frit, Herbst.

2. Elytra with feeble or no tubercles before apex ; tibiae sinuate on inner-side, and thickened above middle.

A A. Body oblong ; tarsi short.

a a. Thorax with three depressions at apex ; all interstices on elytra about equally raised.

Nearly oblong. Black, dull, covered with gray scales ; antennae (except club), apex of tibiae and the tarsi rust-red. Thorax slightly broader than long, very strongly constricted before apex, with three depressions in front, in the central one of which the deep central furrow ends ; elytra with alternate interstices slightly broader, all nearly equally convex, with a tubercle before apex. L. 1¾ l. Rare.

B. brevis, Gyll.

b b. Thorax without apical depressions ; alternate interstices on elytra distinctly more raised.

Oblong. Black ; tibiae rust-red ; covered with brown scales, checkered with gray and white, sides and central line of thorax covered with whitish scales, elytra with middle of disc spotted with white. Thorax closely granulate, deeply constricted toward apex, anterior margin raised ; elytra with deep punctured striae, interstices convex, alternate ones raised ; tibiae slightly curved.   L. 1 l.   Not very common.   *B. lutulosus*, Gyll.

B B. Body elongate ; tarsi long.

Elongate.   Black ; base of antennae and tibiae rust-red ; covered closely with grayish scales, two spots on thorax and others on elytra bare.   Thorax nearly as long as broad, rather deeply constricted toward apex, anterior margin raised, narrowed behind ; elytra with distinct punctured striae, alternate interstices raised.   L. 1¼ l.   Rather common.   *B. tempestivus*, Herbst.

II. Sides of thorax strongly rounded after apical constriction.

Oblong.   Black ; base of antennae and tibiae red-brown, tarsi pitch-brown ; closely covered with gray scales, unequally on elytra, thorax with two points in front and two spots at base bare.   Thorax almost half as broad again as long, constricted toward apex, anterior margin raised, sides rounded, narrowed behind, finely punctured ; elytra with punctured striae, interstices convex.   L. 1½ l.   Moderately common.   *B. limosus*, Gyll.

b. Antennae inserted far before middle of rostrum ; third tarsal joint distinctly broader than second.

I. Femora and tarsi black.

Similar to *B. lutulentus*, but with antennae (except funiculus), femora and tarsi black, second joint of latter almost transverse ; thorax rather shorter, somewhat rounded at sides, narrowed behind, disc more strongly punctured (somewhat wrinkled), posterior angles rather obtuse ; alternate striae on elytra less raised ; rostrum punctured at apex.   L. 1¾ l.   Cambridge.   *B. nigritarsis*, Th.

II. Femora and tarsi rust-red.

1. Elytra with interstices flat, equal, suture raised behind, apex produced.

Elongate.   Black ; legs brown-red ; with gray scales. Rostrum moderately long and thick, curved ; thorax scarcely so long as broad, closely and finely punctured, with an indistinct central furrow, sides nearly straight ; elytra narrowed at apex, with fine punctured striae, with a distinct callosity before apex ; tibiae long, curved.   L. 1¾ l.   Rare.   *B. lutosus*, Gyll.

2. Elytra with alternate interstices distinctly broader and raised, suture not raised behind, apex not produced.

Oblong. Black; tibiae rust-red; variegated with gray scales; each elytron with a white spot behind middle on third interstice. Rostrum moderately long and curved; forehead furrowed; thorax closely punctured, constricted in front, sides nearly straight; elytra with indistinct punctured striae, suture and alternate interstices raised, with a feeble callosity before apex. L. 1¼ l. Common.

B. lutulentus, Gyll.

B. Funiculus of antennae with six joints.

a. Body oblong.

Oblong. Black; antennae (except club) and legs red-brown; closely covered with whitish scales, thorax with four dorsal spots (anterior ones indistinct), elytra with diffusely placed markings of brown scales. Rostrum rather long, curved; thorax broader than long, deeply constricted toward apex, closely and finely punctured, with feeble central furrow, sides nearly straight; elytra with distinct, but diffusely and feebly punctured striae, inner interstices rather convex, outer ones flat. L. 1¾ l. Rare.

B. inceratus, Gyll.

b. Body elongate.

Narrow, cylindrical. Black; antennae and tibiae red-brown; closely covered with gray scales. Thorax almost longer than broad, sides and base straight; elytra scarcely broader than thorax, with fine striae, strongly compressed before apex, without callosity before apex; tarsi elongate, scarcely shorter than tibiae, second and third joints equal in breadth, former not transverse. L. 1¼—1½ l. Not common.

B. cylindrus, Payk.

Hydronomus.

Black; funiculus of antennae, tibiae and tarsi yellow-brown; under-side rather closely and evenly, upper-side more scantily covered with grayish or yellowish-white scales, placed more closely on sides and central line of thorax, apex of elytra and several spots on their disc. Rostrum rather short, nearly straight; thorax impressed on each side; interstices on elytra flat; tibiae curved. L. 1¼—1½ l. Common.

H. alismatis, Marsh.

Mecinus.

A. Thorax without white central longitudinal line ; tibiae black.

a. Rostrum somewhat long and thin, much curved.

Black, rather shiny ; with scanty gray pubescence. Thorax very closely punctured ; elytra with punctured striae, interstices punctured somewhat in rows ; femora with a small sharp tooth. Rostrum of female longer than that of male and impunctate at apex. L. 1½ l. Common.

M. pyraster, Herbst.

b. Rostrum short, robust, nearly straight, punctured.

Black, rather shiny ; with scanty gray pubescence ; posterior margin of thorax and sides of breast closely covered with yellowish scales. Thorax rounded, very closely punctured ; elytra with not very deep punctured striae, interstices indistinctly punctured ; femora almost without tooth. L. 1⅓ l. Scarce.   *M. collaris*, Germ.

B. Thorax with white central line ; tibiae red-brown.

Brown, rather dull ; antennae, tibiae and tarsi redbrown ; sides and suture of elytra usually reddish ; with gray pubescence ; forehead, sides and central line of thorax and sides of elytra closely covered with whitish pubescence. Thorax not very closely punctured ; elytra with punctured striae, interstices punctured somewhat in rows ; femora not toothed. L. 1⅓ l. Not common. *M. circulatus*, Marsh.

Erirhinus.

A. Femora not toothed.

a. Hinder pairs of tibiae straight, with a small tooth at apex.

I. Elytra pitch-brown or blackish.

1. Rostrum with longitudinal striae.

Black, dull ; antennae and legs pitch-brown ; with close brown pubescence ; sides of metasternum with white scales ; each elytron generally with a whitish spot behind middle. Rostrum longer than head and thorax, curved ; thorax closely punctured in wrinkles, with an indistinct central ridge, sides moderately rounded ; elytra more than half as long again as together broad, with indistinct striae (deeper toward suture), interstices rather convex. L. 3—3½ l. Common. *E. scirpi*, Fab.

2. Rostrum punctured or smooth.

A A. Thorax with a yellow spot on each side; length 4 lines.

Black, dull; antennae and legs pitch-brown; pubescence brown; thorax with a curved streak of yellow hairs on each side; each elytron with a more or less distinct whitish spot behind middle. Rostrum as long as head and thorax, curved, shiny, diffusely punctured; thorax closely punctured in wrinkles, with central ridge, sides rounded; elytra with indistinct striae, interstices nearly flat; anterior tibiae toothed on inner-side. Not common.

E. bimaculatus, Fab.

B B. Thorax without yellow spots; length 2—2½ lines.

a a. Thorax without raised central line.

Black, shiny, bare; antennae and legs red-brown. Rostrum as long as head and thorax, curved; thorax almost as long as broad, diffusely and strongly punctured, central line impunctate but not raised, sides slightly rounded; elytra oblong-ovate, with deep punctured striae, interstices flat, extremely finely punctured. L. 2½ l. Rare.

E. aethiops, Fab.

b b. Thorax with a fine raised central line.

Pitch-brown, dull; antennae and legs (except often femora) brown-red; pubescence yellowish-gray, scanty. Rostrum as long as head and thorax, curved, punctured diffusely in front, more closely and partly in rows behind; thorax closely punctured; elytra scarcely one-third longer than together broad, with moderately strong punctured striae, interstices nearly flat. L. 2—2¼ l. Common.

E. acridulus, Lin.

II. Elytra reddish.

Elongate. Red-brown; antennae and legs rust-red; covered with a gray crust, with scattered, erect, whitish bristles. Rostrum almost as long as head and thorax, curved, wrinkled; thorax scarcely broader than long, distinctly punctured, sides slightly rounded; elytra with distinct punctured striae, alternate interstices somewhat raised. L. 1½ l. Not uncommon.   *E. pillumus*, Gyll.

b. All tibiae curved, with a strong tooth at apex.

    I. Each elytron with a round white spot behind middle; length 2¼ lines.

Oblong. Pitch-brown; rostrum, antennae and legs rust-red; closely covered with gray scales (on thorax more scantily), elytra indistinctly variegated with brown. Rostrum closely punctured; thorax much rounded at sides, closely and finely punctured. L. 2⅓ l. Not uncommon.   *E. festucae*, Herbst.

    II. Elytra without round light spot behind middle; length 1½—1¾ lines.

     1. Rostrum nearly smooth.

Elongate, parallel-sided. Pitch-brown; rostrum, antennae and legs rust-red; covered with grayish scales, scantily on disc of thorax and space at base of elytra round scutellum, closely on rest of body. L. 1½—1¾ l. Common.

         E. nereis, Payk.

     2. Rostrum punctured and striate (more strongly in male than in female).

Elongate. Pitch-black; rostrum, antennae and legs rust-red; covered with yellowish-gray scales, scantily on disc of thorax, in spots on elytra and closely on sides of thorax, breast and abdomen. Rostrum thicker than in *E. nereis*, elytra broader. L. 1¾ l. Not common.

         E. scirrhosus, Gyll.

B. Femora with a tooth before apex.

    a. Rostrum as long as half the body, thin, much curved, strongly striate; legs long and thin, anterior pair (especially of male) much longer than the others.

Pitch-black; antennae and legs rust-red; pubescence unequal, gray; elytra variegated with reddish and black, with numerous markings of gray pubescence. Thorax transverse, rapidly narrowed in front, sides strongly rounded; femora with a small tooth. L. 2½—3 l. Common.   *E. vorax*, Fab.

    b. Rostrum not or scarcely longer than head and thorax together; legs somewhat thick, anterior pair only slightly elongate.

I. Elytra more or less distinctly spotted.

1. Sides of thorax much rounded.

A A. Rostrum somewhat striate at base only.

Brown; rostrum black; antennae and legs yellow-brown; pubescence unequal, gray; elytra variegated with gray and brown. Rostrum long, thin, moderately curved; thorax much broader than long, closely punctured, a little narrowed behind, somewhat constricted in front; femora thick, clubbed, strongly toothed. L. 2—2½ l. Not common.

E. tremulae, Payk.

B B. Rostrum with five ridges reaching nearly to apex.

a a. Thorax gradually dilated from apex to middle, broadest in middle.

Elongate. Pitch-black; antennae and legs red-brown; pubescence scanty, gray; elytra checkered with red-brown. Rostrum long; thorax short, closely punctured; tooth on femora moderately strong. L. 2¼ l. Scarce; on Aspens.

E. costirostris, Gyll.

b b. Thorax rather suddenly dilated behind apex, broadest before middle.

Elongate. Pitch-black or brown; thorax with anterior margin narrowly, and posterior margin broadly paler; elytra lighter or darker brown, variegated with dark spots; antennae reddish, with club black; legs reddish or yellow. L. 2—2½ l. Common; on Sallows and Willows.

E. maculatus, Marsh.

E. Silbermanni, Wenck, is described as being larger and broader than *E. costirostris,* with a shorter and thicker rostrum and the spots on the elytra uniform; from *E. maculatus* it is said to differ in being less convex, generally a little larger, with thorax less closely punctured and femora stronger.

2. Sides of thorax not, or not much rounded.

A A. Rostrum not or scarcely longer than thorax.

a a. Rostrum wrinkled longitudinally; thorax much broader than long.

Oblong. Pitch-brown; antennae and legs brown-red;

pubescence gray, rather close, spotted; elytra spotted with black, with a white-haired callosity before apex. Rostrum only slightly curved; thorax closely punctured, indistinctly ridged in front, sides slightly rounded; tooth on femora pointed. L. 2¼ l. Rare.   *E. affinis*, Payk.

b b. Rostrum closely punctured; thorax only slightly broader than long.

Oblong. Black; antennae and legs rust-red; pubescence grayish, spotted; elytra spotted with brown, with a white-haired callosity before apex. Rostrum straight, shorter and thicker than in *E. affinis*; thorax closely punctured, scarcely ridged, sides nearly straight; tooth on femora pointed. L. 2 l. Rather common.

E. validirostris, Gyll.

B B. Rostrum longer than thorax.

a a. Prosternum not emarginate at apex; thorax more finely punctured.

Rather narrow. Pitch-brown; antennae and legs reddish-yellow; pubescence gray, spotted; elytra yellow-brown, with dark spots and a white-haired callosity before apex. Rostrum striate; thorax short, diffusely punctured; tooth on femora pointed. L. 1½ l. Moderately common.

E. taeniatus, Fab.

b b. Prosternum somewhat emarginate at apex; thorax more strongly punctured.

A a. Rostrum more or less striate.

A 1. Thorax a little broader than long.

Oblong, rather shiny. Red, with scanty, fine, pale pubescence; head, rostrum and breast black; elytra scarcely variegated behind. Rostrum longer than thorax, nearly straight, deeply striate; thorax a little broader than long, scarcely narrower at apex than at base, sides somewhat rounded, strongly punctured, with central line impunctate. L. 1¼ l. Not common.   *E. salicis*, Walt.

B 1. Thorax as long as broad.

Rather narrow. Pitch-black; legs reddish-yellow; pubescence gray, spotted; breast closely covered with white pubescence. Thorax longer than in *E. taeniatus*, sides somewhat rounded. L. 1½ l. Rather common.

E. salicinus, Gyll.

B b. Rostrum wrinkled or punctured.

A 1. Head black.

Brown-red ; head (with or without all or part of rostrum) and under-side black ; pubescence unequal, gray ; elytra with dark spots near suture. Rostrum as long as head and thorax, rather thick, wrinkled, moderately curved ; thorax narrowed in front, closely punctured, sides scarcely rounded. L. 1½ l. Rare. *E. majalis*, Payk.

B 1. Head yellow-red.

a 1. Body oblong.

Oblong. Yellow-red; under-side pitch-black ; pubescence gray, on thorax close at sides, scanty on disc ; elytra with a large pitch-brown common spot at base, rather closely covered throughout with short pubescence. Rostrum curved, striate and punctured from base to middle ; apex reddish ; thorax somewhat broader than long, narrowed in front, closely punctured. Narrower than *E. pectoralis*, with rostrum longer and more curved, sides of thorax more rounded. L. 1½ l. Rather common. *E. agnathus*, Boh.

b 1. Body oblong-ovate.

Oblong-ovate. Yellow-red ; breast brown or (blackish) ; apex of rostrum brown ; pubescence gray, unequal on elytra, which are variegated with brown and gray. Rostrum longer than head and thorax, moderately curved, striate and wrinkled at base ; thorax punctured, sides moderately rounded. L. 1½ l. Common. *E. pectoralis*, Panz.

II. Elytra not spotted.

Reddish-yellow, rather shiny ; breast brown ; pubescence gray, scanty throughout. Rostrum nearly as long as half the body, slightly curved, shiny, punctured at base ; thorax closely punctured, with feeble central furrow. L. 2½ l. Moderately common. *E. tortrix*, Lin.

Grypidius.

Pitch-black ; breast, sides of thorax and elytra, and apex of latter closely covered with white and brownish-gray scales ; each elytron with a white spot in middle. Alternate interstices on elytra raised, third, fifth and seventh with callosity behind middle. L. 2¾—3 l. Common.

G. equiseti, Fab.

Procas.

Oblong-ovate. Black, dull; antennae and legs reddish-pitch-brown; pubescence scanty, gray, matted; elytra spotted with gray and brown. Thorax very closely punctured, with an indistinct raised central line; elytra with distinct punctured striae. L. 3 l. Rare. *P. Steveni*, Gyll.

HYLOBIIDES.

A. Anterior coxae placed slightly apart.

Rostrum generally slender, cylindrical; antennae inserted near middle of rostrum, short; eyes oval; elytra with a callosity before apex; second abdominal segment much longer than third and fourth together, separated from first by a curved suture; tarsal claws free.

Pissodes, Germ.

B. Anterior coxae contiguous.

a. Rostrum cylindrical.

Rostrum more or less robust; antennae inserted toward apex of rostrum, moderately long; eyes elongate; elytra with a callosity before apex; second abdominal segment as long as next two together, separated from first by a straight or angular suture; tarsal claws free.

Hylobius, Germ.

b. Rostrum somewhat angular, flat and finely ridged above.

Rostrum moderately robust; antennae inserted toward apex of rostrum, moderately long; eyes oval; elytra scarcely broader than thorax; second abdominal segment a little longer than next two together, separated from first by a straight suture; tarsal claws free. *Lepyrus*, Germ.

Pissodes.

A. Base of thorax slightly sinuate, posterior angles scarcely prominent.

Lighter or darker pitch-brown; scantily covered with yellowish scales, here and there in spots; elytra with a narrow band (formed of small spots) behind middle and a spot (usually double) before middle of yellowish-white or yellow scales. Elytra with punctured striae, punctures of dorsal striae large, deep, oblong. L. 3½—4 l. Moderately common. *P. pini*, Lin.

B. Base of thorax deeply sinuate, posterior angles prominent, acute.

a. Elytra with white spot before middle.

Similar to *P. pini* but with markings paler; large punctures of dorsal striae oblong-quadrate. L. 3—4 l. Scarce.
P. notatus, Fab.

b. Elytra without white spot before middle.

Oblong-ovate, not very convex. Brown-red, with white scales; a transverse row of small spots on thorax, the scutellum and an interrupted band near middle of elytra closely covered with white scales. Punctures on thorax farther apart than in *P. notatus* and not confluent. L. 2 l. Rare. *P. piniphilus*, Herbst.

Hylobius.

Pitch-black, dull; covered with narrow yellowish-gray scales, placed more closely here and there; elytra with from two to four irregular bands of yellow spots. Thorax deeply punctured in wrinkles, constricted in front, with central ridge; elytra with chains of punctures, interstices flat, strongly wrinkled. L. 4—6 l. Common.
H. abietis, Lin.

Lepyrus.

Black; covered with brownish-gray scales and hairs; thorax with a white line on each side; elytra with a small whitish spot before apex. Thorax dilated behind, with a somewhat indistinct central ridge. L. 4½—5 l. Very rare. *L. binotatus*, Fab.

CLEONIDES.

A. Rostrum cylindrical, scrobes generally commencing more or less far from apex.

a. Body cylindrical; scrobes not meeting beneath.

Rostrum variable; first two joints of funiculus of antennae often long, equal to each other or not; eyes oval, oblong or elongate; thorax transverse or not, more or less regularly

conical, slightly tubular in front, base bisinuate, central
lobe short and narrow; elytra not or scarcely broader than
thorax, apex of each sometimes produced into a point.
                                                    Lixus, Fab.

b. Body oblong or short oval; scrobes nearly always
meeting beneath.

Rostrum variable in length, rather robust; first two joints
of funiculus of antennae slightly elongate, rest very short,
gradually enlarged to club; eyes elongate; thorax transverse, strongly narrowed and abruptly tubular in front, base
nearly straight, with a distinct central lobe; elytra a little
broader than thorax.   *Larinus*, Germ.

B. Rostrum angular; scrobes commencing near or at apex.

a. Rostrum at most as long as head, robust, scarcely
curved, scrobes meeting beneath.

First two joints of funiculus of antennae slightly elongate, nearly equal; eyes elongate; thorax transverse,
strongly narrowed in front, base bisinuate, central lobe
rather broad; elytra not broader than thorax.
                                            Rhinocyllus, Germ.

b. Rostrum longer than head, moderately robust, slightly
curved; scrobes not meeting beneath.

I. Posterior tarsi not elongate, more or less broad,
generally spongy beneath, third joint rarely not
longer than second.

Eyes elongate; thorax bisinuate at base; elytra not or
scarcely broader than thorax.   *Cleonus*, Schoenh.

II. Posterior tarsi elongate, not spongy beneath,
third joint distinctly longer than second.

Eyes elongate; thorax with base bisinuate, ocular lobes
broad and prominent; elytra not broader than thorax.
                                        Bothynoderes, Schoenh.

Lixus.

A. Scape of antennae as long as whole of funiculus.

a. Apex of each elytron produced, pointed, divergent.

I. Points at apex of elytra nearly as long as thorax.

Elongate. Black, with gray pubescence, powdered with
greenish-yellow. Thorax longer than broad, only a little

narrowed in front, extremely finely and closely punctured in wrinkles, anterior margin with long fringe behind eyes; elytra with indistinct punctured striae. L. 6—7 l. Not common. *L. paraplecticus*, Lin.

II. Points at apex of elytra short.

Broader than *L. paraplecticus*. Black, with gray pubescence, powdered with greenish-yellow; sides of thorax and two streaks on its middle rather lighter. Elytra with rows of distinct punctures. L. 6—7½ l. Rare.

L. turbatus, Fab.

b. Apex of elytra separately rounded.

I. Rostrum longer than thorax, moderately curved, strongly punctured; length 6—6½ lines.

Elongate. Black; antennae brown-red; with scanty gray pubescence. Thorax almost longer than broad, somewhat narrowed in front, roughly wrinkled; elytra with distinct rows of punctures, impressed rather deeply on shoulders, slightly behind scutellum. Not common.

L. angustatus, Fab.

II. Rostrum shorter than thorax, straight, wrinkled, ridged; length 3½—4 lines.

Elongate. Black; with gray pubescence, powdered with brownish; orbit of eyes with yellow hairs; thorax with a yellow line on each side; elytra with grayish-white spots. Thorax as long as broad, somewhat narrowed in front, roughly punctured; elytra with fine rows of punctures, only slightly impressed at base. Not common.

L. bicolor, Ol.

B. Scape of antennae scarcely longer than first three joints of funiculus.

Elongate. Black; antennae and tarsi rust-red; with gray pubescence, powdered with yellow; thorax with four yellow lines. Rostrum almost as long as thorax, closely punctured, indistinctly channeled at base; thorax conical, with a deep transverse impression in front, closely and finely granulate; elytra scarcely impressed at base, separately rounded at apex, with close and fine punctured striae, interstices closely granulate. L. 3—4 l. Rare.

L. filiformis, Fab.

Larinus.

Elongate. Black; with gray pubescence; elytra spotted with gray. Rostrum shorter than thorax, curved, cylindrical; thorax very finely wrinkled; elytra with punctured striae, interstices broad, flat, finely wrinkled. L. 3½ l. Not common.   *L. carlinae,* Ol.

Rhinocyllus.

Oblong-ovate. Black, dull; antennae and tarsi pitch-brown; pubescence matted, yellowish-gray, in spots (especially on elytra). Rostrum impressed, with a distinct interrupted central ridge; punctuation close. L. 2½ l. Not common.   *R. latirostris,* Latr.

Cleonus.

Oblong. Black; pubescence rather close, gray; thorax with five white lines, (outer ones curved); elytra with two not very distinct oblique bands bare. Rostrum with three deep furrows throughout, about equal to each other in breadth; thorax narrowed in front, diffusely granulate, with slight central ridge in front and depression at base. L. 5½—7½ l. Common.   *C. sulcirostris,*

Bothynoderes.

A. First joint of funiculus of antennae longer than second.

a. Vertex of head distinctly ridged.

Black; pubescence gray; thorax with grayish lines; elytra with two oblique bands bare. Rostrum with a central ridge and furrow on each side, ridge split between base of antennae, enclosing an oblong depression; thorax slightly narrowed in front, punctured in wrinkles, with central ridge in front and depression at base; elytra with punctured striae, with irregular depressions, narrowed behind (more so in female than in male). L. 5½—6 l. Not common.   *B. nebulosus,* Lin.

b. Vertex of head scarcely ridged.

Black; pubescence close, grayish-white; elytra with two bands and apical callosity bare. Vertex of head scarcely ridged; elytra with a distinct humeral callosity and with deep punctured striae, somewhat pointed at apex. L. 4—5 l. Rare.   *B. glaucus,* Fab.

B. Second joint of funiculus of antennae longer than first.

Black; pubescence close, whitish; disc of thorax and on elytra two bands and a spot before apex bare. Rostrum somewhat narrowed at apex, with two furrows, the central ridge forked in front; elytra separately pointed at apex, points divergent. L. 4—4½ l. ´ Rare. *B. albidus,* Fab.

HYPERIDES.

A. Funiculus of antennae with six joints.

First joint of funiculus of antennae larger and longer than second; rostrum rather robust, scrobes commencing at some distance from apex; other parts as in *Hypera.*

Limobius, Schoenh.

B. Funiculus of antennae with seven joints.

a. Antennae inserted in or near middle of rostrum.

Antennae thin, scape reaching to anterior margin of eyes or farther, first two joints of funiculus nearly equal; rostrum at most moderately robust, scrobes generally commencing close to mouth; thorax straight at base and apex, sides more or less rounded; second abdominal segment shorter than next two together. *Hypera,* Germ.

b. Antennae inserted near apex of rostrum.

Antennae rather thin, scape not reaching beyond anterior margin of eyes, first two joints of funiculus nearly equal; rostrum moderately robust; thorax transverse, straight at base, sides rounded in front; second abdominal segment at least as long as next two together. *Alophus,* Schoenh.

Limobius.

A. Second interstice on elytra immaculatate behind.

Brown; antennae and legs yellow-red; rostrum rust-red; covered with slightly metallic, grayish, yellow and brown scales; thorax with three lines of light scales; elytra with whitish and dark spots and long, upright, white and black hairs, hinder half of suture with whitish scales. Thorax half as broad again as long, sides strongly rounded; elytra about one-third longer than together broad. L. 1¼ l. Not common. *L. dissimilis,* Herbst.

B. Elytra with a common dark spot behind extending across second and third interstices.

Oblong-ovate. Black; antennae and legs yellow-red; covered with pale-brown scales; thorax with a line of pale scales on each side; elytra with a small spot on each side at base and a transverse common spot in middle dark brown, also a common whitish spot on hinder part of suture. Thorax much broader than long, sides moderately rounded; elytra with very fine punctured striae. L. $1\frac{1}{4}$— $1\frac{1}{2}$ l. Rather common.　　　　　　　　　　*L. mixtus,* Boh.

Hypera.

A. Second joint of funiculus of antennae much longer than third joint.

a. Length 3 lines or more.

I. Thorax more narrowed behind than in front.

1. Second row of punctures on elytra nearly parallel with suture at base; length $3\frac{1}{2}$—$3\frac{3}{4}$ lines.

Winged. Black; antennae and tibiae brown-red; closely covered with brown and gray hairlike scales, central line and sides of thorax, sides of elytra and the under-side closely covered with grayish scales. Rostrum thick, shorter than thorax, latter much broader than long; elytra with rather deep punctured striae, alternate interstices with lighter scales, with rows of black or brown velvety spots and diffusely-placed, erect, white hairs. L. $3\frac{1}{2}$—$3\frac{3}{4}$ l. Common.　　　　　　　　　　　　　　　*H. punctata,* Fab.

2. Second stria on elytra divergent from suture at base; length 3 lines.

Similar to *H. punctata,* but with rostrum thinner and scarcely shorter than thorax, latter more rounded at sides; elytra with scales lighter in colour, a spot at shoulder and another in middle of side margin closely covered with whitish scales. L. 3 l. Moderately common.

　　　　　　　　　　　　　　　H. fasciculata, Herbst.

II. Thorax evenly narrowed before and behind.

Apterous. Black; antennae (except club) red; with gray or bright-brown pubescence; thorax with three rather lighter lines; elytra sometimes with brown spots. Rostrum moderately curved; thorax convex, closely punctured;

elytra nearly double as long as together broad, with punctured striae, interstices slightly convex. L. 3 l. Rare.

H. elongata, Payk.

III. Thorax with sides nearly parallel from base to beyond middle and thence narrowed in front.

Apterous. Black; closely covered with ochre-yellow scales and also with fine bristles (on elytra placed in rows); thorax with two broad brown longitudinal lines; elytra without spots. L. 3 l. Rare. *H. arundinis*, Fab.

b. Length under 3 lines.

I. Elytra spotted.

1. Sides of thorax only slightly rounded.

A A. Striae on elytra equal in depth.

a a. First joint of funiculus of antennae half as long again as second; rostrum nearly straight.

Black; covered with grayish scales; thorax with two indistinct dark longitudinal lines; elytra with small brown spots. Rostrum thin, nearly straight; thorax scarcely broader than long. L. 2—2¼ l. Common.

H. rumicis, Lin.

b b. First joint of funiculus of antennae only slightly longer than second; rostrum somewhat curved.

Black; covered with grayish scales; thorax with two indistinct dark longitudinal lines; elytra closely checkered with black. Rostrum shorter than in *H. rumicis*, somewhat curved; thorax rather more rounded at sides, but scarcely broader than long. L. 2½ l. Common.

H. pollux, Fab.

B B. Alternate striae on elytra deeper.

Black; covered with bronze-gray, rather shiny scales; thorax and elytra with white lines. Rostrum short, rather thick, somewhat curved; thorax scarcely broader than long. L. 2½ l. Moderately common. *H. julini*, Sahlb.

2. Sides of thorax strongly rounded.

A A. Base of suture of elytra without common dark spot.

a a. Elytra variegated throughout.

A a. Thorax rather flat.

Oblong. Black; antennae and tibiae reddish-brown; with pubescence and gray scales; thorax with two brown longitudinal lines; elytra with brown lines on disc and diffusely-placed black spots throughout. Thorax rather flat, sides strongly rounded in front. L. 2¼ l. Not common. *H. tigrina*, Boh.

B b. Thorax convex.

Oblong. Black; antennae rust-red; closely covered with gray pubescence and scales; thorax with two broad brown longitudinal lines; elytra with numerous dark spots (often confluent), hinder half of suture not or more scantily spotted with black than front part. Rostrum short, only slightly curved; thorax convex. L. 2½—2¾ l. Not uncommon. *H. suspiciosa*, Herbst.

b b. Elytra with a large, dark spot on disc of each.

Oblong-ovate. Black; antennae and tibiae brown-red; closely covered with gray scales; thorax with two broad longitudinal brown lines; elytra with a large, elongate dark spot on disc of each and usually with other dark markings toward scutellum. Rostrum moderately long, curved, thorax rather flat. L. 2¼ l. Rather common.

H. plantaginis, De G.

B B. Base of suture of elytra with a common dark spot.

a a. Thorax narrower at apex than at base; length 2¾ lines.

Black; antennae and tibiae rust-red; covered with gray scales; thorax with two brown longitudinal lines, sometimes with three pale greenish lines; elytra with white lines, spotted with black, base of suture with a common oblong, sometimes toothed brown spot. L. 2¾ l. Common.

H. murinus, Fab.

b b. Thorax nearly as broad at apex as at base; length 2—2¼ lines.

Oblong-ovate. Black; antennae and tibiae rust-red; with gray scales and short pubescence; thorax with two brown longitudinal lines and a dark point on each side before middle; elytra with indistinct black points, base of suture (until beyond middle) with a triangular dark common spot. Rostrum moderately long, slightly curved; thorax rather flat. L. 2—2¼ l. Common.

$H.$ *variabilis*, Herbst.

II. Elytra with uninterrupted light and dark lines.

Narrow. Black; head and thorax with brown scales and hairs; latter with three white lines, central one continued on to head; elytra gray, with suture brown from before middle to apex, dilated into spots in places, second interstice brown in front, whitish behind, third interstice from middle and fifth from before middle backward with brown lines meeting toward apex. Thorax rather broader than long, sides moderately rounded. L. 2½ l. Common.

$H.$ *polygoni*, Lin.

B. Second joint of funiculus of antennae not much longer than third joint.

a. Elytra without dark spot on disc.

I. Thorax nearly twice as broad as long.

Pitch-black; antennae and tibiae rust-red; closely covered with gray or yellowish hairlike scales; thorax with two broad longitudinal dark lines; elytra with a row of whitish hairs on interstices, apex of suture spotted with white and brown. L. 1¾ l. Rare. $H.$ *meles*, Fab.

II. Thorax not much broader than long.

Pitch-black; antennae and legs brown-red; thorax with three green lines; elytra covered with green hairlike scales. L. 1½ l. Common. $H.$ *nigrirostris*, Fab.

b. Elytra with a dark spot on disc of each.

Black, closely covered with grayish-metallic scales; antennae and legs rust-red; thorax with two brown lines; elytra with suture spotted with brown, base of third interstice and a spot on middle of disc of each brown, nearly bare. Thorax not much broader than long, sides moderately rounded. L. 2 l. Moderately common.

$H.$ *trilineata*, Marsh.

Alophus.

Oblong-ovate. Black; closely covered with gray or brown scales; each elytron with two light spots, anterior one variable in size, placed before middle on disc, hinder one placed before apex, large and attached to corresponding one on other elytron. Rostrum with a deep central furrow throughout; thorax closely punctured, with a distinct central furrow on anterior half; elytra with indistinctly punctured striae. L. 2½—4 l. Common. *A. triguttatus*, Fab.

MOLYTIDES.

A. Posterior coxae globular.

   a. Elytra convex, oval.

Rostrum elongate, slender, curved, cylindrical, scrobes commencing in or near middle, curved; antennae short, rather robust, funiculus with seven joints; thorax transverse, convex, sides rounded.   *Trachodes*, Germ.

   b. Elytra flat, oblong oval.

Rostrum considerably longer than head, moderately robust, slightly curved, angular, more or less convex and ridged above, scrobes reaching mouth or nearly so, straight; antennae moderately long, not very robust, funiculus with seven joints; thorax at least as long as broad, not very convex, central line ridged, sides more or less rounded.

                                                 Plinthus, Germ.

B. Posterior coxae oval, transverse.

   a. Tibiae feebly hooked at apex; scrobes confluent behind.

Mandibles very short, slender, cutting in front; funiculus of antennae with seven joints; thorax nearly straight at sides, base rounded, apex straight, without trace of ocular lobes; scutellum absent; elytra scarcely broader at base than thorax, shoulders rectangular; tarsi narrow, spongy beneath.   *Liosomus*, Steph.

   b. Tibiae strongly hooked at apex; scrobes not confluent behind.

Mandibles pincherlike, convex in front, toothed on inner-side; funiculus of antennae with seven joints; thorax ovoid, base and apex truncate, ocular lobes very feeble; scutellum curvilinear triangular; elytra a little broader at base than thorax, shoulders rounded; tarsi rather broad, first two joints furrowed and spongy at margins only.

                                               Molytes, Schoenh.

Trachodes.

Pitch-brown; rostrum, antennae and legs lighter; upper-side with erect black scales ; under-side, sides of thorax and some bandlike spots on elytra with yellowish-white or gray scales. Elytra with deep striae, alternate interstices apparently raised, suture soldered ; femora toothed. L. 1½ l. Not uncommon. *T. hispidus*, Lin.

Plinthus.

Elongate. Black; antennae and legs reddish-brown; scales gray. Rostrum indistinctly punctured in wrinkles, with fine central ridge ; thorax oblong, roughly punctured, flat, ridged ; elytra with deeply punctured striae, suture and alternate interstices raised ; femora with sharp tooth. L. 4 l. Scarce. *P. caliginosus*, Fab.

Liosomus.

A. Femora toothed.

Nearly ovate. Black, shiny, bare ; antennae brown-red, legs red-brown ; sides of breast with white pubescence. Rostrum moderately curved, finely punctured ; thorax deeply and rather closely punctured, with an indistinct smooth central line ; elytra with rows of large round punctures, interstices broad, flat, with a row of very fine punctures and of short and exceedingly fine bristles. L. 1¼ l. Common. *L. ovatulus*, Clairv.

The variety *collaris*, Rye, is smaller, with femora dark at apex, thorax often red, less closely punctured, antennae and legs rather longer, tooth on femora feebler.

B. Femora not toothed.

a. Thorax shiny.

Oblong-ovate. Black, shiny, bare ; antennae pale rust-red ; legs pitchy-red. Similar to *L. ovatulus*, but narrower, with rostrum less curved, more deeply punctured, sides of thorax less curved, antennae (especially scape) longer and inserted nearer apex of rostrum, punctures on elytra larger, forming rows, but apparently not placed in impressed lines, under-side more strongly and diffusely punctured. L. 1¼ l. Rare. *L. oblongulus*, Boh.

b. Thorax dull.

Ovate, rather short. Black, bare ; antennae bright rust-red, club short, darker; femora pitch-black, tibiae and tarsi rust-red. Rostrum less stout than in *L. ovatulus*, not so distinctly thickened before insertion of antennae and much more strongly punctured (punctures forming irregular longitudinal furrows), antennae and legs more slender, anterior tibiae almost straight; broader than *L. oblongulus*, elytra with striae more distinct and shoulders more prominent, antennae inserted farther from apex of rostrum. Thorax strongly and very closely punctured. L. (including rostrum) 1¼ l.   Rare.   *L. troglodytes*, Byc.

Molytes.

A. Femora with a sharp tooth.

Black; two lateral transverse spots on thorax and its posterior margin with yellow scales; elytra unspotted or with only a few spots (not impressed). Thorax finely punctured, sides slightly rounded. L. 4½—5½ l. Rather common.   *M. coronatus*, Latr.

B. Femora indistinctly toothed.

Black; sides of thorax and the elytra with many impressed spots covered with yellow scales. Thorax rather deeply punctured, sides rounded. L. 7—10 l. Not common.   *M. germanus*, Lin.

RHYPAROSOMIDES.

Rostrum longer than head, somewhat angular, curved, scrobes straight, not turning on to under-side; funiculus of antennae with six joints; eyes oval; thorax without ocular lobes, prosternum not impressed before anterior coxae, scutellum absent.   *Orthochaetes*, Germ.

Orthochaetes.

Oblong.   Pitch-brown; antennae and legs red-brown; head and thorax with rough recumbent gray hairs, latter about as long as broad, punctured in wrinkles; elytra with deep punctured striae, suture and alternate interstices raised, with a row of erect, whitish-yellow bristles. L. 1¼ l. Moderately common.   *O. setiger*, Germ.

BYRSOPSIDES.

Forehead more or less impressed; rostrum rather long,

slightly curved; sternal channel superficial; antennae short, moderately robust, funiculus with seven joints, first of them elongate, rest short, seventh separate from club; thorax cylindrical or nearly so; elytra broader than thorax, oblong, or slightly oval; truncate surface at apex of posterior tibiae with outer margin not reflexed.

Gronops, Schoenh.

Gronops.

Ovate. Black; closely covered with grayish-white scales; elytra with a large brown spot, pointed toward shoulders. Thorax with several depressions; elytra with hinder half of suture and alternate interstices strongly raised, fifth with a callosity before apex. L. 1¾ l. Moderately common. *G. lunatus*, Fab.

LEPTOPSIDES.

Rostrum a little longer than head, rather robust, rounded at angles, finely ridged above; antennae moderately long, rather stout, funiculus with seven joints, first two joints knotted at apex, first longer; thorax a little broader than long, sides slightly rounded, ocular lobes scarcely distinct; scutellum absent; elytra not broader than thorax; tarsi rather long, spongy beneath, claws soldered.

Tropiphorus, Schoenh.

Tropiphorus.

A. Thorax with a very fine raised central line; elytra with hinder part of suture and alternate interstices feebly raised.

Nearly ovate. Black; antennae and tibiae red-brown; strewn with brownish-coppery scales. Elytra with indistinct punctured striae. L. 3 l. Not common.

T. mercurialis, Fab.

B. Thorax with a strongly raised central line; elytra with hinder part of suture and alternate interstices strongly raised and bearing a row of whitish bristles.

Oblong-ovate. Black; antennae and tibiae red-brown; strewn with brownish-coppery scales. Elytra with regular, but not deep, punctured striae, fifth interstice with a strong callosity behind. L. 2½ l. Not common.

T. carinatus, Müll.

OTIORHYNCHIDES.

A. Metasternum elongate; intercoxal process relatively narrow, with its sides not parallel.

Head prolonged behind eyes; rostrum at most as long as head, slightly dilated at apex, angles rounded, apex more or less emarginate; antennae variable in length and stoutness; thorax transverse, sides rounded in middle, base and apex truncate; scutellum distinct; elytra broader than thorax, parallel-sided for two-thirds of their length, shoulders obtuse; legs rather long, tarsal claws soldered; second abdominal segment shorter than next two together; separated from first by a nearly straight suture.

Phyllobius, Germ.

B. Metasternum very short; intercoxal process broad, parallel-sided.

a. Antennae at most moderately long, more or less stout.

I. Tarsal claws free.

1. Thorax regularly cylindrical.

Rostrum as in *Trachyphloeus;* antennae stouter, first joint of funiculus very large; elytra oval, a little broader than thorax, and strongly emarginate at base.

Cathormiocerus, Schoenh.

2. Thorax not cylindrical.

Rostrum as long as, or a little longer than, and as broad as head, separated from it by a transverse furrow, slightly curved, nearly parallel-sided, angular, apex slightly emarginate; scrobes lateral, deep, slightly curved, reaching eyes; antennae inserted in middle of rostrum; short, robust, first two joints of funiculus (especially first) elongate; thorax transverse, sides generally much rounded, base and apex truncate; scutellum absent or very small; elytra oval, scarcely broader than thorax, slightly emarginate at base.   *Trachyphloeus*, Germ.

II. Tarsal claws soldered.

1. Scrobes lateral.

A A. Scrobes deflexed, or if not, body with scales.

a a. Body with scales or bristles, scrobes generally entirely deflexed.

Scrobes lateral, generally turned under eyes, but not touching them ; rostrum longer and less thick than in *Barypeithes* and *Omias* ; thorax always broader than long ; femora moderately clubbed, not toothed.

Platytarsus, Schoenh.

b b. Body hairy or bare, scrobes deflexed on lower part.

Scrobes lateral, deflexed on to under surface, the part near margin deeper ; rostrum generally as long as broad ; thorax generally broader than long ; femora clubbed, rarely toothed. Barypeithes, Duv.

B B. Scrobes not deflexed ; body without scales.

Scrobes lateral, ending far from eyes ; rostrum thick, rather long ; thorax somewhat broader than long ; femora not toothed. Omias, Schoenh.

2. Scrobes on upper surface of rostrum.

A A. Head striate at sides and beneath.

Rostrum as long as, and narrower than head, angular-scrobes deep, widened behind, nearly reaching eyes ; antennae inserted near middle of rostrum, more or less short and robust, first two joints of funiculus elongate, nearly equal ; thorax rounded at sides and base, truncate at apex ; scutellum absent ; tarsi short, narrow. Caenopsis, Bach.

B B. Head not striate at sides.

Rostrum not longer than, and almost as broad at base as head, angles nearly rounded, pterygia strongly divergent, apex truncate ; antennae inserted near apex of rostrum, rather long and robust, first two joints of funiculus elongate, almost equal ; thorax transverse, sides and base rounded, apex truncate ; scutellum rarely distinct ; body closely covered with scales, without short, recumbent hairs.

Peritelus, Germ.

b. Antennae long and slender.

Rostrum at least as long as head, nearly horizontal, robust, parallel-sided and angular or nearly rounded, or entirely rounded at base, with pterygia more or less divergent, apex more or less emarginate, upper-side often with a ridge (bifid at apex) ; scrobes straight, deep and visible from above in front, effaced behind ; antennae inserted before middle of rostrum, first two joints of funiculus elongate, nearly equal or second longer than first; thorax generally almost as long as broad, sides rounded, base and apex truncate; scutellum absent or very small; tarsi moderately broad, spongy beneath, claws free; second abdominal segment considerably longer than either of next two, separated from first by an angular suture.

Otiorhynchus, Germ.

Phyllobius.

<u>*A*</u>. Femora toothed.

  a. Body with scales.

    I. Upper-side with oblong scales.

      1. Third to seventh joints of funiculus of antennae conical; length $3\frac{1}{2}$—4 lines.

        A A. Thorax indistinctly constricted in front; scutellum semi-oval, apex rounded.

Oblong. Black, somewhat hairy, not very closely sprinkled with grayish hairlike scales; antennae and legs generally red or brown-red. Thorax narrowed in front. Not uncommon.   *P. calcaratus,* Fab.

        B B. Thorax distinctly constricted in front; scutellum triangular, apex pointed.

Very similar to *P. calcaratus,* but with the scales green or blue-green. Common.   *P. alneti,* Fab.

      2. Third to seventh joints of funiculus of antennae nearly globular; length $2\frac{1}{2}$—3 lines.

Oblong. Black; covered with oblong coppery scales, usually lighter coloured on alternate interstices on elytra; scutellum with white scales; antennae and legs generally reddish-yellow. Antennae thick; thorax distinctly consticted in front; scutellum triangular, apex pointed. Common.   *P. pyri,* Lin.

    II. Upper-side with round scales.

      1. Elytra with long, erect hairs; antennae entirely pale brownish-yellow.

Oblong. Black; closely covered with green scales, with some silvery reflection, with long hairs; antennae, tibiae and tarsi pale brownish-yellow. Antennae rather thick. L. $2\frac{1}{2}$—3 l. Common.   *P. argentatus,* Lin.

2. Elytra with very short hairs, scarcely projecting beyond scales ; antennae yellow-red, with apex of scape and the club black.

Oblong. Black ; closely covered with green or blue-green scales ; hairs short, white ; antennae as above ; legs black, with apex of tibiae and tarsi red-brown or brown. Rostrum flat above, with central channel. L. 2—2¼ l. Common. *P. maculicornis*, Germ.

b. Body without scales.

Elongate. Black ; elytra brown, with or without blackish margin ; antennae and legs brown-red ; with rather long, gray pubescence. L. 2—2¼ l. Common.

P. oblongus, Lin.

B. Femora not toothed.

a. Elytra closely covered with scales.

I. Breast and abdomen closely covered with green scales.

Similar to *P. uniformis* but with thorax longer, less dilated at sides, flatter and sometimes ridged in middle ; elytra longer. L. 2—2¼ l. Common. *P. pomonae*, Ol.

II. Breast and abdomen diffusely covered with green scales.

Black ; antennae, tibiae and tarsi red-yellow ; upper-side closely covered with pale green dull scales. Antennae rather thick ; rostrum moderately long. L. 1¾—2¼ l. Common. *P. uniformis*, Marsh.

b. Elytra without scales.

Oblong-ovate. Black, shiny ; antennae, tibiae and tarsi brown-yellow ; femora pitch-brown ; sides of thorax and the breast covered with green scales. Antennae rather thick ; rostrum short ; thorax closely punctured, sides moderately rounded ; elytra with rather deep punctured striae, interstices scarcely convex. L. 1½—2 l. Common. *P. viridicollis*, Fab.

Cathormiocerus.

A. Scrobes not reaching eye.

Oblong-ovate. Pitch-black ; antennae and legs brown ; covered with gray scales, on upper-side scantily in middle, closely at sides, on under-side not very closely. Rostrum with fine central channel; thorax scarcely broader than long, closely punctured; elytra with regular, distinct punctured striae, interstices rather flat.   L. 1¼ l.   Rare.

C. socius, Boh.

B. Scrobes reaching eye.

More robust, flatter, darker, and much more strongly punctured than C. socius, with head wider; eyes more prominent; rostrum thicker, with central impressed line less distinct; scrobes when viewed from the front less open, when viewed laterally more regular, rounded on upper and angulated on lower edges, reaching eye, smooth and shiny; scape of antennae shorter and stouter, less abruptly angulated at base above constricted portion on side next eye, but more angulated on outer-side, funiculus and club broader and shorter; thorax less transverse, more strongly rounded at sides and more narrowed at base, more distinctly punctured; elytra with more coarsely punctured striae, bristles on interstices black, tessellation more conspicuous in fresh specimens; knees and outer-side of tibiae more or less pitchy.   L. 1½—1¾ l.   Rare.

C. maritimus, Rye.

Trachyphloeus.

A. Thorax without central furrow.

   a. Anterior tibiae with two not very strong teeth.

   I. Elytra with distinct striae, interstices equally raised.

Ovate. Black; antennae and legs brown-red; closely covered with gray scales and on upper-side scantily with short white bristles. Rostrum flat above; antennae inserted at base of scrobes; thorax much broader than long, with a broad, deep, curved, transverse impression near apex, sides strongly rounded; elytra with distinct punctured striae, interstices level, each with a row of white bristles; teeth on anterior tibiae short.   L. 1¼ l.   Weston-super-Mare.

T. laticollis, Boh.

   II. Elytra with indistinct striae, alternate interstices more strongly raised.

Ovate. Pitchy, covered with a grayish-yellow crust; antennae and legs dusky yellow. Rostrum scarcely furrowed; alternate interstices of elytra a little raised, with a diffusely-placed row of bristles.　L. 1—1¼ l.　Scarce.

T. alternans, Gyll.

b. Anterior tibiae with three pointed teeth, central one forked at apex.

Black ; antennae and legs red-brown ; with a close crusted covering, and on upper-side scantily set with bristles. Rostrum flat above ; thorax broadest before middle, constricted toward apex ; elytra with indistinct striae, alternate interstices sometimes slightly raised ; anterior tibiae as above.　L. 1—1¼ l.　Scarce.

T. spinimanus, Germ.

B. Thorax with a more or less distinct central furrow.

a. Apex of anterior tibiae with a strong tooth on outerside and another (bifid at apex) in front ; teeth small in female.

Pitch-black ; antennae and legs lighter ; upper-side (especially on elytra) spotted with whitish scales ; interstices on elytra with a row of thick, erect yellowish scales. Rostrum with a broad furrow ; thorax almost double as broad as long, strongly constricted before apex, sides rounded ; elytra short ovate, with distinct striae.　L. 1¼ l. Common.　　　　　　　　　　　<i>T. scabriculus</i>, Lin.

b. Apex of anterior tibiae with six minute spines at apex.

Ovate.　Black ; antennae and legs brown, with brown or grayish-brown scales ; interstices on elytra with a row of erect, rather thick bristles. Rostrum with a broad furrow ; thorax somewhat constricted toward apex, usually with an oblong depression on each side of disc, sides strongly rounded ; elytra almost ovate, with distinct striae. L. 1⅓—1½ l.　Rather common.　　　　<i>T. scaber</i>, Lin.

c. Anterior tibiae not toothed at apex.

I. Elytral bristles stout, clubbed.

1. Second abdominal segment scarcely as long as next two together, separated from first by a straight suture.

Ovate.　Reddish-brown ; elytra, antennae and legs paler ; with erect, white scales throughout. Rostrum with a central furrow ; sides of thorax much dilated. L. 1½ l. Scarce.　　　　　　　　　　　　<i>T. aristatus</i>, Gyll.

2. Second abdominal segment longer than next two together, separated from first by a curved suture.

Similar to *T. aristatus* but with thorax not so wide, elytra more elongate, elytral bristles not quite so strong; differing from *T. squamulatus* in having the eyes larger, thorax more rounded at sides and bristly, second abdominal segment rather longer, scrobes less horizontal, with upper margin less sharply defined. L. 1½ l. Southsea beach.

            T. myrmecophilus, Seid.

II. Elytral bristles feeble.

Ovate. Reddish-brown; elytra, antennae and legs rather lighter; not very closely covered with grayish scales; upper-side with white bristles. Rostrum and thorax with feeble central furrow, latter scarcely constricted in front, sides only moderately dilated; elytra with more or less distinctly punctured striae. L. 1½ l. Scarce.

            T. squamulatus, Ol.

Platytarsus.

A. Rostrum without central furrow.

Black-brown or brown; antennae and legs pale yellow-brown; pubescence gray; sides of thorax closely covered with round whitish scales; elytra with long erect bristles. Head punctured; rostrum and thorax without central furrow, latter finely and very closely punctured in wrinkles; elytra globular, with deep punctured striae. L. 1¼ l. Rather common.   *P. echinatus*, Bons.

B. Rostrum with a central furrow or depression.

Ovate, convex. Black or dark brown; antennae and tarsi rust-red; tolerably closely covered with gray or brownish scales, and on upper-side scantily with whitish bristles. Rostrum with a central furrow or depression; thorax closely and finely punctured; elytra nearly ovate, with deep, somewhat indistinctly punctured striae. Narrower than *P. echinatus*, less convex, thorax longer, more rounded at sides. L. 1½ l. Rare.   *P. setulosus*, Boh.

Barypeithes.

A. Forehead not impressed or with only a small feeble impression; thorax without lateral impressions behind middle.

a. Thorax deeply and rather closely punctured, sides strongly rounded.

Oblong-ovate. Pitch-brown, shiny; antennae and legs yellow-red; with gray pubescence. Rostrum impressed, distinctly and closely punctured; elytra oblong-ovate, with deep punctured striae. L. 1¾ l. Rare. *B. pellucidus,* Boh.

b. Thorax deeply and diffusely punctured, sides moderately rounded.

Oblong-ovate. Pitch-brown or brown, rather shiny; antennae and legs red-yellow; with very scanty short, grayish pubescence. Rostrum dilated at apex, impressed, indistinctly punctured; elytra ovate, with deep punctured striae. L. 1½ l. Common. *B. brunnipes,* Ol.

B. Forehead with a deep furrow, continued to apex of rostrum; thorax with a transverse impression on each side behind middle.

Oblong. Black, shiny; antennae and legs yellow-red, with scanty grayish pubescence; rostrum rust-red at apex, closely punctured. Thorax broad, deeply and rather closely punctured, sides strongly rounded; elytra oblong, with deep punctured striae. L. 1½ l. Not uncommon.

B. sulcifrons, Boh.

Omias.

Ovate. Pitch-brown; antennae and legs yellow-red; elytra red-brown, strewn with short pale bristles. Head indistinctly punctured, vertex almost smooth; thorax rather broader than long, indistinctly and not closely punctured, sides moderately rounded; elytra ovate, with deep punctured striae. L. 1½ l. Not common. *O. mollinus,* Boh.

Caenopsis.

A. Rostrum with a broad and deep longitudinal furrow.

Black-brown; legs brown; closely covered with dark brown and gray scales; sides of thorax with gray scales and small, rather depressed bristles; sides of elytra with gray, disc with brown scales; posterior femora with a white ring; rostrum short, with a broad and deep furrow ending in a depression on head; thorax dilated behind middle, with a narrow raised central line; elytra with punctured

striae, second interstice broader than first. L. 2½ l. Very
rare.   *C. fissirostris*, Walt.

B. Rostrum with a fine central furrow.

Ovate. Black; antennae and legs rust-red, with gray
scales, elytra with lighter markings, rather closely covered
throughout with short, thick, white bristles. Thorax finely
punctured, with indistinct central furrow, sides strongly
rounded; elytra globular, with deep punctured striae. L.
1¼ l. Rather common.   *C. Waltoni*, Boh.

Peritelus.

Oblong-ovate. Black; antennae and legs reddish-pitch-
brown; covered and somewhat checkered with brown, gray
and whitish scales. Forehead with a small depression
between eyes; rostrum with a fine central furrow; thorax
deeply and diffusely punctured, sides slightly rounded,
scales white at sides, brown on disc; elytra oblong-ovate,
with fine and not very deep punctured striae, sides with
white scales, disc with brown and gray, mixed with white
lines. L. 2½—3½ l. Rare.   *P. griseus*, Ol.

Otiorhynchus.

A. Femora not toothed.

   a. Anal segment of male regularly striate longitudinally;
      second joint of funiculus of antennae longer than first.

   I. Antennae with joints of funiculus longer.

Oblong-ovate. Black, rather shiny; legs red-brown;
elytra with small tufts of depressed gray pubescence; rostrum emarginate, with feeble central ridge; thorax narrow,
very finely shagreened (as also head), sides moderately
rounded before middle; elytra closely wrinkled, narrowed at
apex, with very indistinct striae. L. 5½ l. Moderately
common.   *O. tenebricosus*, Herbst.

   II. Antennae with joints of funiculus shorter, in
      female broader than long.

Oblong. Black, somewhat shiny, with scanty gray pubescence; legs red, knees and tarsi black; rostrum ridged
in middle; thorax oblong, closely punctured and granulate;
elytra with rather remote and tolerably deep punctured
striae, interstices transversely wrinkled, apices jointly

rounded. L. 4½—5½ l. Not common.

O. haematopus, Schoenh.

b. Anal segment of male punctured or at most scratched, even or with a shallow depression.

I. Eyes more or less approximated, forehead narrow, not or not much arched transversely.

1. All interstices evenly raised.

A A. Elytra without rows of bristles on interstices.

a a. Upper-side bare.

Elongate-ovate. Black, shiny. Thorax closely shagreened at sides, diffusely punctured on disc; elytra oblong-ovate, with distinct punctured striae, inner interstices feebly wrinkled, outer ones granulate. L. 5—5½ l. Rare.

O. ebeninus, Schoenh.

b b. Upper-side pubescent.

A a. Breast with longer and closer pubescence than upper-side.

Black, rather shiny; legs reddish-brown. Rostrum ridged; head and thorax very finely shagreened, latter as long as broad, rounded at sides; elytra ovate, obtuse, with very indistinct striae, interstices granulate. L. 3½—4 l. Common.

O. atroapterus, De G.

B b. Pubescence on breast not closer or longer than on upper-side.

Black; antennae and legs reddish-brown; under-side, head and thorax scantily covered with pubescence; elytra closely covered with gray and brown pubescence, clouded. Rostrum punctured in wrinkles; thorax rather broader than long, closely granulate, with a short and fine central ridge; elytra short ovate, with deep punctured striae (hidden by scales). L. 3 l. Not uncommon.

O. raucus, Fab.

B B. Elytra with rows of bristles on interstices.

a a. Rostrum furrowed.

Oblong-ovate. Black: antennae, tibiae and tarsi brown-red; with scanty brown pubescence. Forehead with deep furrow; thorax closely covered with rounded tubercles; elytra with strong punctured striae, interstices closely

wrinkled and granulate. L. 3½ l. Moderately common.
O. scabrosus, Marsh.

b b. Rostrum not furrowed.

Oblong-ovate. Black, slightly shiny; antennae and legs
rust-red ; with feeble gray pubescence. Rostrum flat, not
ridged ; thorax closely covered with obtuse tubercles, sides
moderately rounded ; elytra with moderately deep punctured striae, interstices rather convex, wrinkled, sides and
apex with sharp tubercles. L. 2¾—3¼ l. Rather common.
O. ligneus, Ol.

2. Alternate interstices on elytra more raised.

Ovate. Brown-red ; antennae and legs paler ; variegated
with gray and white scales. Rostrum not ridged ; thorax
rather broader than long, finely and rather closely granulate, sides rounded ; elytra with rows of eye spots, suture
and alternate interstices raised, with rows of bristles. L.
2⅓ l. Not common.   *O. septentrionis*, Herbst.

II. Eyes placed apart, forehead broad and strongly
arched transversely.

1. Upper-side pubescent.

Nearly ovate. Black ; antennae and legs reddish-brown ;
with short, scanty gray pubescence ; rostrum with central
ridge ; thorax short, closely granulate, sides rounded ; elytra
ovate, with shallow punctured striae, interstices finely
wrinkled. L. 2¾ l. Not common.   *O. maurus*, Gyll.

2. Upper-side bare.

Ovate. Black, shiny, bare ; antennae and legs pitchy ;
rostrum punctured in wrinkles ; elytra punctured diffusely
in front and on disc, very closely at sides and apex, impressed striae scarcely perceptible. L. 3—3½ l. Rather
common.   *O. blandus*, Gyll. (*monticola*, Sharp's Cat.)

B. Femora toothed.

a. Elytra closely covered with scales and punctured
with eye-spots.

Oblong-ovate. Black-brown, dull ; antennae and legs
lighter ; closely covered and variegated with brown and
gray scales. Rostrum scarcely furrowed ; thorax almost as
long as broad, coarsely granulate, sides rounded ; elytra
with striae, punctured with eye-spots ; femora with an indistinct obtuse tooth. L. 3—3¼ l. Common. *O picipes*, Fab.

b. Elytra not, or scantily covered with scales, not punctured with eye-spots.

I. Antennae slender, second joint of funiculus much longer than first.

Oblong. Black; antennae pitch-brown; elytra with scattered spots of grayish-yellow scales. Rostrum with central furrow; thorax at least as long as broad, rather closely granulate, with a very feeble central furrow; elytra with deep furrows. L. 4½ l. Common. *O. sulcatus*, Fab.

II. Antennae more or less thick, second joint of funiculus not much longer than first.

1. Length 4—5½ lines.

Black; with gray scales. Rostrum with central ridge; thorax granulate; elytra closely and finely granulate, with punctured striae toward sides. Scarce. *O. ligustici*, Lin.

2. Length 2—3 lines.

A A. Antennae thickened toward apex.

a a. Rostrum longitudinally wrinkled; second joint of funiculus of antennae rather longer than first.

Oblong-ovate. Black, dull; with scanty gray pubescence. Forehead and rostrum longitudinally wrinkled; thorax broader than long, closely granulate, sides moderately rounded; elytra rather flat on disc, with moderately strong punctured striae, interstices convex, granulate in rows, with short white bristles. L. 2½ l. Rather common.
O. rugifrons, Gyll.

b b. Rostrum with a longitudinal ridge and a shallow furrow on each side; second joint of funiculus of antennae almost shorter than first.

Ovate. Black, dull. Pubescence closer than in *O. rugifrons*; rostrum and forehead more punctured in wrinkles, but with the punctures more evident. Thorax rather more finely granulate; elytra with rather finer punctured striae, granulation of interstices not in distinct rows. L. 2¼—2½ l. Rare. *O. ambiguus*, Schoenh.

B B. Antennae not thickened toward apex.

a a. Body black.

Short ovate. Black ; antennae and legs red-brown ; with scanty gray pubescence. Rostrum punctured in wrinkles ; thorax almost globular, closely wrinkled, with short central ridge ; elytra shiny, with rather fine punctured striae, dorsal interstices flat, nearly smooth, lateral ones finely wrinkled ; femora strongly toothed. L. 2—2$\frac{1}{3}$ l. Common.   *O. ovatus*, Lin.

b b. Body rust-red.

Rust-red ; antennae and legs red-yellow. Similar to *O. ovatus*, but with antennae and legs shorter and less robust; thorax less strongly tuberculated, disc only feebly wrinkled; elytra longer, narrower and more parallel-sided, with pubescence shewing traces of spots; rostrum narrower. L. 2 l. Not uncommon.   *O. muscorum*, Bris.

BRACHYDERIDES.

A. Anterior margin of thorax with a flattened tuft of stiff hairs on each side behind eyes.

Rostrum at most as long as head, thick, parallel-sided or nearly so, angular, flat, often with a central ridge or furrow, apex triangularly emarginate or nearly entire ; scrobes short, curved, widened behind, usually not reaching eyes ; antennae rather long and slender, thorax longer than broad ; scutellum elongate triangular ; elytra broader (sometimes only slightly) at base than thorax, shoulders callous or obliquely rounded ; truncate surface at apex of posterior tibiae very oblique, acutely triangular ; its outer margin more or less reflexed, forming a cavity ; third joint of posterior tarsi much broader than first and second.
                                                 Tanymecus, Schoenh.

B. Anterior margin of thorax without stiff hairs behind eyes.

    a. Truncate surface at apex of posterior tibiae (on which is inserted the tarsi) with outer margin not reflexed.

        I. Elytra broader at base than thorax, shoulders angular.

           1. Tarsal claws soldered.

                A A. Scrobes meeting beneath.

Head slightly elongate, reversed conical ; rostrum shorter and a little narrower than head, rather thick, parallel-sided, somewhat rounded at angles, slightly convex above, apex nearly entire, or moderately emarginate ; scrobes narrow, abruptly curved ; antennae usually slender, scape generally reaching much beyond posterior margin of eyes ; scutellum very small, curvilinear triangular ; elytra somewhat broader at base than thorax ; legs moderately long, tarsal claws soldered at base. *Polydrosus*, Germ.

B B. Scrobes not meeting beneath.

Similar to *Polydrosus* but with rostrum very short, parallel-sided, angular, flat above, apex slightly emarginate or entire ; antennae moderately long, scape often curved, reaching a little beyond posterior margin of eyes.

Metallites, Germ.

2. Tarsal claws free.

Rostrum at most as long as, and a little narrower than head, parallel-sided or gradually narrowed in front, slightly and narrowly emarginate at apex, flat and usually finely furrowed above, scrobes rather deep, abruptly curved ; antennae short or moderately long, not very robust, scape reaching at most to posterior margin of eyes ; scutellum very small ; elytra considerably broader than thorax ; tarsal claws free. *Sitones*, Germ.

II. Elytra not broader at base than thorax, shoulders rounded.

1. Eyes only moderately prominent.

A A. Rostrum considerably narrower than head, rounded at angles, truncate or scarcely emarginate at apex.

Scrobes very short, rather deep and broad, moderately curved, widened and effaced behind ; antennae long and slender, scape as long as funiculus ; thorax transverse, slightly rounded at sides ; legs long, tarsal claws soldered. *Eusomus*, Germ.

B B. Rostrum nearly as broad as head, angular, moderately emarginate at apex.

Scrobes rather deep, short, abruptly curved, ending far from eyes ; antennae long and slender, scape reaching more or less beyond eyes, very slender ; thorax generally transverse, sides slightly rounded ; legs long, tarsal claws soldered. *Sciaphilus*, Schoenh.

2. Eyes strongly prominent.

Rostrum sometimes separated from forehead by a transverse furrow, very short, wedge-shaped, at other times continuous with forehead, longer and parallel-sided, always angular, flat above, more or less emarginate at apex ; scrobes deep, well defined, curved ; antennae moderately long, not very robust, scape generally reaching beyond posterior margin of eyes ; thorax generally very short, sides more or less rounded ; legs moderately long, tarsal claws soldered.

Strophosomus, Schoenh.

b. Truncate surface at apex of posterior tibiae with outer margin reflexed, forming a cavity.

I. Rostrum rounded at angles, distinctly narrower than head, more or less dilated at apex, scutellum generally distinct.

1. Tarsal claws free.

Rostrum distinctly longer than head, flat or slightly convex above, often with central furrow, more or less emarginate and fringed at apex ; scrobes deep, curved, usually reaching eyes ; antennae moderately long, rather robust, scape reaching middle of eyes, elytra scarcely broader at base than thorax, shoulders rectangular or obtuse.   *Barynotus*, Germ.

2. Tarsal claws soldered.

Rostrum scarcely longer than head, a little convex and finely ridged above, truncate and fringed at apex, scrobes superficial or moderately deep, rather wide, curved, ending far from eyes ; antennae rather long, not very robust, scape reaching beyond posterior margin of eyes ; elytra broader at base than thorax, shoulders very obtusely angular or rounded.   *Liophloeus*, Germ.

II. Rostrum angular, as broad (or nearly so) as head, parallel-sided or slightly narrowed in front ; scutellum very small or absent.

Rostrum at least as long as head, flat, generally with a fine central ridge, often separated from forehead by a transverse furrow, more or less emarginate and fringed at apex, scrobes deep, curved, a little widened behind ; antennae moderately long, rather robust, scape reaching nearly to posterior margin of eyes ; tarsal claws soldered.

Cneorhinus, Schoenh.

Tanymecus.

Elongate. Black ; closely covered with scales and hairs, on upper-side brownish-gray, on sides and under-side grayish-white. Rostrum with an indistinct impression ; elytra with distinct punctured striae. L. 4—4½ l. Moderately common. *T. palliatus*, Fab.

Polydrosus.

A. Scape of antennae reaching beyond eyes.

a. Last five joints of funiculus of antennae nearly round.

Elongate. Black; antennae and legs red ; upper-side brown ; sides of thorax, sides (sometimes dilated obliquely from shoulder) and apex of elytra, also a band (curved backward) behind their middle, and the under-side with grayish-white scales. Thorax as long as broad, sides almost straight; femora not toothed. L. 1¾—2½ l. Common.
P. undatus, Fab.

b. Last five joints of funiculus of antennae obconical.

I. Elytra without line of paler scales at sides.

1. Scales on elytra evenly distributed.

A A. Upper-side without hairs ; femora indistinctly toothed.

Oblong. Black ; antennae rust-red, apex of separate joints and the club brown ; covered with rather dull green scales, without pubescence. Forehead flat, sometimes impressed in middle ; rostrum a little shorter than head. L. 2½—3 l. Very rare. *P. planifrons*, Schoenh.

B B. Upper-side with hairs; femora not toothed.

a a. Vertex of head without prominences ; scales dull.

Elongate. Black; antennae and legs reddish-yellow ; closely covered with green scales and upright, brownish pubescence. Rostrum very short ; thorax somewhat broader than long. L. 2⅓—3 l. Rare.
P. flavipes, De G.

b b. Vertex of head with a large transverse prominence on each side above eyes; scales shiny.

Oblong, narrower than *P. flavipes.* Black ; antennae

and legs reddish-yellow ; closely covered with green scales and upright, brown pubescence. Rostrum very short ; forehead flat ; thorax much broader than long ; elytra with very fine punctured striae. L. 2¼ l. Common.

P. pterygomalis, Schoenh.

2. Interstices on elytra with dark, bare spots.

Oblong. Black; antennae reddish at base ; strewn throughout with round, gray or greenish, shiny scales, except on some bare spots on elytra. Elytra with punctured striae ; femora toothed. L. 2—2½ l. Common.

P. cervinus, Lin.

II. Elytra with a line of paler scales at sides.

1. Rostrum much shorter than head.

Oblong. Black ; antennae, tibiae and tarsi red-yellow ; with scanty brown pubescence and with pale greenish scales placed diffusely on disc of thorax and elytra, closely at sides and on third interstice on latter. Thorax transverse, sides scarcely rounded; elytra with distinct punctured striae ; femora toothed. L. 2—2¼ l. Not common.

P. chrysomela, Ol.

2. Rostrum scarcely shorter than head.

Oblong. Black ; antennae, tibiae and tarsi yellow-brown ; with feeble pubescence and with scattered yellowish scales, sides of thorax and elytra with a streak of white scales. Thorax transverse, rounded at sides ; femora toothed. L. 2½—3 l. Not uncommon.

P. confluens, Steph.

B. Scape of antennae reaching at most to posterior margin of eye.

a. Scales round.

Oblong. Black ; antennae and legs reddish-yellow ; covered with round, dull, green scales, without pubescence. Rostrum scarcely shorter than head ; forehead slightly impressed ; elytra with moderately strong punctured striae, interstices flat ; posterior femora toothed. L. 2—3½ l. Rare.

P. sericeus, Schall.

b. Scales hairlike.

Black ; antennae and legs red ; covered with narrow golden or reddish scales, breast with whitish scales. Elytra large, strongly convex behind; posterior femora feebly toothed. L. 3½—4 l. Rather common. *P. micans*, Fab.

Metallites.

Pitch-black; antennae and legs reddish-yellow-brown; rather closely covered with gray, slightly metallic hairs. Thorax as long as broad, sides moderately dilated; scutellum much broader than long, truncate; elytra with deep punctured striae, interstices double as broad as punctures; anterior femora with pointed tooth, punctuation close. L. 1⅓—1¾ l. Rather common. *M. marginatus*, Steph.

Sitones.

A. Thorax truncate in front and behind, moderately dilated in middle, sides rather angular than rounded; elytra convex, narrowed in posterior third-part and somewhat pointed at apex.

a. Elytra elongate, more than twice as long as together broad.

Elongate. Black; covered with brown scales, thorax with three indistinct lines, elytra with a broad common spot (spotted with black on each side) of gray scales; femora with whitish scales and a brown band before apex. Thorax almost broader than long, punctured in wrinkles, with central furrow, sides rounded in middle: elytra nearly two and a half times as long as thorax, without erect bristles. L. 3 l. Rather common. *S. griseus*, Fab.

b. Elytra not elongate.

I. Eyes flat, or, if somewhat prominent thorax distinctly longer than broad.

1. Upper-side with gray, brown, or greenish scales; interstices on elytra more or less convex.

A A. Anterior margin of thorax not raised.

a a. Thorax as broad as long.

Oblong. Black; base of antennae and tibiae rust-brown; closely covered on upper-side with gray or brown, on underside with whitish-gray scales. Thorax with three lines of light scales; elytra with or without alternate interstices lighter, sometimes partly checkered with black. Thorax as long as broad, very finely punctured, broadest in middle;

elytra with regular punctured striae. L. 2—2½ l. Rather common.   *S. flavescens*, Marsh.

b b. Thorax longer than broad.

Oblong. Black ; antennae, tibiae and tarsi red-yellow ; covered with gray pubescence, thorax with three lines, elytra with badly defined markings white. Thorax a little longer than broad, closely and very finely punctured, sides slightly rounded ; elytra with fine punctured striae. L. 2¼ l. Rare.   *S. longicollis*, Schoenh.

B B. Anterior margin of thorax raised.

Black ; scantily covered with gray or greenish scales ; thorax with sides and sometimes a central line pale; elytra with suture and a streak along each pale ; legs and base of antennae dull red. L. 2—2¼ l. Common.

S. suturalis, Steph.

2. Upper-side with scarcely metallic scales ; interstices on elytra quite flat.

Oblong. Black ; scape of antennae, tibiae and tarsi red-yellow, clothed above with scarcely metallic scales, beneath with grayish ones. Eyes rather flat; elytra with punctured striae, interstices flat. On the average rather longer than *S. suturalis*, less cylindrical, and not so convex ; eyes a little less prominent (although more so than in *S. hispidulus*) and when viewed from the side visibly narrower in their transverse diameter and more elliptical in outline ; elytra with feebler punctured striae. L. 2 l. Scarce.

S. ononidis, Sharp.

II. Eyes moderately prominent ; thorax as long as broad.

Oblong. Black ; antennae, tibiae and tarsi rust-red ; upper-side scantily covered with coppery scales, thorax with three coppery lines ; elytra more or less distinctly variegated with white ; under-side with very fine whitish pubescence, with a lateral streak (abbreviated behind) of silvery or bluish scales. Rostrum somewhat hollowed out at apex, with marginal line raised and a very short central ridge ; forehead with deep furrow ; thorax finely punctured ; elytra with distinct punctured striae. L. 1¼ l. Common.   *S. sulcifrons*, Thunb.

B. Thorax very slightly dilated at sides ; elytra not very convex, elongate and almost parallel-sided ; eyes tolerably prominent.

a. Upper-side with short, silky, recumbent pubescence.

I. Sides of elytra parallel.

1. Thorax as long as broad.

Oblong. Black ; base of antennae, tibiae and tarsi red-yellow ; unequally covered with white, silvery or slightly greenish scales ; thorax with three badly defined light lines ; elytra with scales placed more closely on suture, side and some streaks on disc. Thorax about as broad as long, sides more rounded than in *S. lineellus*, less deeply punctured ; elytra with distinct punctured striae. L. 1½ l. Common. *S. tibialis*, Herbst.

2. Thorax broader than long.

Oblong-ovate. Black ; antennae (except club) and tibiae red-yellow ; scantily and nearly evenly covered with whitish scales, placed a little more closely toward sides of thorax. Thorax much broader than long, more rounded at sides than in *S. lineellus*, more closely punctured in wrinkles ; elytra with deep punctured striae. L. 1¼—1½ l. Not common. *S. brevicollis*, Schoenh.

II. Elytra narrower at base than in middle.

Oblong. Black ; antennae, tibiae and tarsi rust-red covered with brown scales, with three well defined white lines on thorax, and on elytra sides and a dorsal streak or lines white. Thorax as long as broad or rather longer, deeply punctured, sides somewhat rounded ; elytra with moderately strong punctured striae, with some short bristles behind. L. 1½ l. Rare. *S. lineellus*, Gyll.

b. Elytra with erect hairs behind.

I. Interstices on elytra convex, not all equally raised.

Elongate. Black, scantily covered with gray scales ; antennae and legs red-yellow. Head broad, coarsely punctured, with strong furrow, reaching to middle of rostrum, latter about as long as head and not much narrower ; eyes very prominent ; thorax oblong, slightly dilated at sides, flat, sides more or less closely covered with scales, disc sometimes glabrous ; elytra with deep punctured striae, interstices very narrow, convex, wrinkled, second and fourth distinctly raised and more or less covered with gray and white scales.   L. 2½ l.   Not uncommon.

S. Waterhousei, Wall.

II. Interstices on elytra flat.

Oblong-ovate.   Black; base of antennae, tibiae and tarsi red-yellow; closely covered with gray, brown or brownish-yellow scales; thorax with a fine central line and a broader one on each side lighter; elytra with long, upright white bristles and with black spots.   Thorax as broad as, or somewhat broader than long, deeply and somewhat diffusely punctured, sides nearly straight; scales on elytra placed more closely than in *S. lineellus*, and bristles longer.   L. 1½ —1¾ l.   Common.   *S. crinitus*, Ol.

C. Thorax much rounded at sides; eyes very prominent.

a. Elytra with sides almost parallel and apex regularly rounded.

I. Pubescence on elytra more diffuse and shorter, black, spotted with gray or brown.

Oblong.   Black; covered tolerably evenly with fine, recumbent, gray pubescence, mixed beneath with gray scales.   Head flat, deeply and rather diffusely punctured; rostrum and forehead with a deep furrow; thorax short, broader than long, deeply and rather diffusely punctured, sides rounded, anterior margin raised; elytra twice as long as thorax, with fine punctured striae.   L. 2½—3 l.   Not uncommon.   *S. cambricus*, Steph.

II. Pubescence on elytra closer and longer, gray, nearly unicolorous.

Oblong.   Black, rather shiny; antennae rust-red; covered on upper-side with gray pubescence, elytra more or less indistinctly variegated with brown, on under-side with grayish-white scales.   Forehead with a short furrow; thorax not longer than broad, diffusely punctured, distinctly constricted toward apex, sides slightly rounded; elytra with distinct punctured striae.   L. 2½—2¾ l.   Rare.

S. cinerascens, Fab.

b. Elytra very broad, and wider behind than in front.

Oblong.   Black; scape of antennae, tibiae and tarsi red-yellow; with gray scales; thorax with three light lines.   Head punctured; rostrum and forehead with a very distinct central furrow; thorax convex, scarcely broader than long, deeply punctured, constricted toward apex, sides evenly rounded, anterior margin raised ; elytra two and a half times as long as thorax, with fine punctured striae, interstices flat, with erect bristles behind. L. 3 l. Common.

S. regensteinensis, Herbst.

D. Thorax moderately rounded at sides ; elytra with sides nearly parallel and apex regularly rounded ; eyes only slightly prominent.

a. Elytra oval ; length 3 lines.

Black, closely covered with brown or gray scales ; thorax rather long, finely punctured, posterior angles slightly prominent, with sides and central line pale, and two, four or six varied pale dots on disc ; base of antennae, tibiae and tarsi dull-red. Common. *S. puncticollis*, Steph.

b. Elytra oblong ; length 2 lines.

Black ; antennae, tibiae and tarsi rust-red ; upper-side with grayish-silvery scales ; thorax with three straight lines, elytra with alternate interstices whitish. Thorax broader than long, broadest behind, very finely punctured. Common. *S. lineatus*, Lin.

E. Thorax moderately rounded at sides ; elytra with parallel-sides, apex somewhat pointed ; eyes flat.

a. Elytra with erect bristles.

Oblong-ovate. Black ; under-side closely covered with gray, upper-side with brown scales, thorax with a fine central line and broad, somewhat curved lateral lines of grayish-white scales ; elytra variegated with gray and with rows of erect white bristles. Thorax broader than long, deeply and diffusely punctured, rather strongly dilated in middle of sides ; elytra scarcely half as long again as together broad, with deep punctured striae. L. 1¾—2 l. Common.

S. hispidulus, Fab.

b. Elytra without erect bristles.

I. Forehead with a feeble, narrow furrow.

Elongate. Black, more or less closely covered with coppery and brown, or gray and silvery scales ; antennae red-yellow at base ; base and apex of femora, the tibiae and tarsi red-yellow. Head narrow ; thorax as broad as long, very closely and finely punctured ; elytra elongate, generally indistinctly variegated. L. 2—2¼ l. Scarce.

S. meliloti, Walt.

II. Forehead with a broad, deep furrow.

Elongate. Black; antennae and legs red-brown; under-side closely covered with white, upper-side with brown scales; three feeble lines on thorax, scutellum, sides of elytra and a short streak at base of each lighter, femora with a white ring toward apex. Thorax not longer than broad, closely and moderately deeply punctured, sides rounded in middle; elytra not dilated at sides, with rather indistinct punctured striae. L. 1½—2 l. Common.

S. humeralis, Steph.

Eusomus.

Black; funiculus of antennae red-brown; closely covered with light green, slightly shiny, round scales. Elytra with fine punctured striae, interstices broad, flat, scales on alternate ones often rather paler. L. 2½—3 l. Rare.

E. ovulum, Ill.

Strophosomus.

A. Elytra not constricted toward base, basal border not raised.

a. Elytra with base of suture black.

Black; antennae and legs rust-red; very closely covered and variegated with gray and brownish scales (except base of suture). Rostrum separated from forehead by an impressed line; thorax coarsely punctured, with a narrow, often indistinct central furrow; elytra with distinct punctured striae, interstices with diffusely placed bristles. L. 2—2½ l. *S. coryli*, Fab.

b. Elytra unicolorous.

Very similar to *S. coryli*; thorax punctured in wrinkles, without central furrow; elytra almost oval. L. 1½—2½ l. Common. *S. obesus*, Marsh.

B. Elytra constricted toward base, basal border raised.

a. Upper-side rather closely covered with scales throughout and with upright hairs.

I. Elytra with only very short upright hairs; forehead without central furrow.

Similar to *S. faber* but with base of thorax truncate, whole body evenly and not very closely covered with silver-gray scales, with very short upright hairs on elytra only. L. 1½—2 l. Rather common. *S. retusus*, Marsh.

II. Whole of upper-side with long, upright hairs; forehead with a distinct central furrow.

Black; antennae and legs reddish-brown; covered with gray scales and erect hairs. Thorax very short, bisinuate at base, with four lines of lighter scales; basal margin of elytra projecting as an acute angle at sides. L. 2½ l. Rather common. *S. faber*, Herbst.

b. Upper-side diffusely covered with scales (except at sides) and without upright hairs.

Black, rather shiny; antennae and legs pitch-brown; covered with silvery or coppery scales, placed closely on sides of thorax, base of suture of elytra, a broad streak along their side margin and the under-side, scantily on rest of upper-side. Head coarsely, thorax rather more finely punctured, base of latter nearly straight; elytra with deep punctured striae. L. 2 l. Common. *S. limbatus*, Fab.

Barynotus.

A. Upper-side variegated with brown and gray scales.

Pitch-black; with brown and gray scales, placed rather closely and in spots (especially on elytra). Rostrum with a central furrow and some longitudinal wrinkles beside this; elytra with indistinct punctured striae, alternate interstices somewhat raised, fifth and seventh united in a curve near shoulder. L. 5 l. Rather common.

B. obscurus, Fab.

B. Upper-side with scales unicolorous.

a. Rostrum with one furrow.

Pitch-black; antennae brownish-red; upper-side covered with coppery scales, under-side with scanty gray pubescence. Rostrum deeply and not very closely punctured, with a deep central furrow; thorax with distinct central furrow, disc shallowly, somewhat closely punctured, sides punctured in wrinkles; elytra oblong-ovate, with fine punctured striae, third and fifth interstices raised. L. 3—4 l. Not common. *B. Schoenherri*, Zett.

b. Rostrum with five furrows.

Pitch-black; covered with gray scales. Rostrum with a central furrow and on each side of this two deep, rather smaller furrows ; elytra with deep punctured striae, fifth and seventh united in a curve at a considerable distance from shoulder and more strongly raised (especially behind) than in *B. obscurus.* L. 4—4½ l. Not common.

.   *B. moerens,* Fab.

Liophloeus.

Black, dull ; antennae (except club) red-brown ; closely covered with gray scales (often with metallic reflection) ; elytra checkered with brown spots. Thorax finely sha-greened, with a feeble central ridge ; elytra with rather deep and distinct punctured striae, interstices flat. L. 4—6 l. Common.   *L. nubilus,* Fab.

Cneorhinus.

A. Thorax finely punctured ; elytra with fine punctured striae, interstices broad and flat.

Ovate. Black; covered with brown scales, with lines of gray, sides of elytra with white scales. Rostrum scarcely constricted at base, flat, not channeled ; thorax transverse, sides rounded ; elytra broad, nearly globular, with fine punctured striae and white bristles. L. 2—3 l. Common.
C. geminatus, Fab.

B. Thorax longitudinally wrinkled ; elytra with deep punctured striae, interstices convex.

Ovate. Black; covered with gray scales. Rostrum and thorax narrower than in *C. geminatus,* former somewhat constricted in middle, channeled, divided by an impressed line from forehead, latter with sides slightly rounded ; elytra large, more inflated than in *C. geminatus,* with distinct punctured striae and white bristles ; eyes more prominent. L. 2—3 l. Moderately common.
C. exaratus, Marsh.

ANTHRIBIDAE.

A. Antennae inserted on sides of rostrum.

a. Scrobes forming a broad depression.

I. Eyes entire.

1. Thorax flattened, angular at sides ; eyes with distinct orbit.

Rostrum parallel-sided, somewhat transverse, truncate in front ; antennae scarcely reaching middle of thorax, first two joints longer than rest ; transverse ridge on thorax placed before base, curved, interrupted and near base in middle, rounded obliquely at ends, reaching middle of sides ; first tarsal joint scarcely longer than second.

Platyrhinus, Clairv.

2. Thorax more or less convex, not angular at sides ; eyes without orbit.

Rostrum square or somewhat elongate ; antennae short ; transverse ridge on thorax placed before base ; first tarsal joint much longer than second.   *Tropideres*, Schoenh.

II. Eyes slightly emarginate.

Rostrum much longer than broad, slightly dilated in front, apex strongly triangularly emarginate ; antennae of male as long as body, with first joint very large, second very small, next six elongate, about equal, last three forming an elongate club ; antennae of female scarcely reaching base of thorax, joints two to eight elongate, third a little longer than rest, last three forming a moderately large club; transverse ridge on thorax confused with base, ends turning at a right angle, and reaching middle of sides ; anterior coxae placed rather far apart ; body oblong.

Anthribus, Geoffr.

b. Scrobes forming a long furrow, reaching whole length of sides of rostrum and much approximated beneath.

Rostrum very short, cut obliquely on each side, truncate at apex ; antennae a little longer than head and thorax together, first two joints larger than rest, about equal, next six very short, last three forming an oblong club ; eyes entire ; tarsi short and rather broad, first and second joints about equal ; body short or oblong oval.

Brachytarsus, Schoenh.

B. Antennae inserted on upper surface of rostrum.

Rostrum strongly transverse, very slightly rounded in front, scrobes small, rounded ; antennae of male a little longer than thorax, of female scarcely as long as thorax ;

eyes entire ; transverse ridge on thorax confused with base, not turning on to sides ; first tarsal joint distinctly longer than second ; pygidium partly covered ; body cylindrical.   *Choragus*, Kirby.

Platyrhinus.

Oblong, flat on disc. Black, rather closely covered with gray and brown hairs ; rostrum, forehead, apex of elytra, breast and abdomen closely covered with white or whitish-yellow hairs. Rostrum wrinkled ; thorax broader than long, disc uneven. Not common.   *P. latirostris*, Fab.

Tropideres.

A. Transverse ridge on thorax straight, or with convexity directed backward.

a. Rostrum rather elongate, base somewhat narrow, dilated at apex.

Oblong, almost parallel-sided. Black ; hairy; sprinkled with gray ; rostrum and a large spot at apex of elytra, under-side and part of legs closely covered with white hair. Eyes approximated. L. 2—3 l. Rare.
    T. albirostris, Herbst.

b. Rostrum short, scarcely dilated at apex.

Oblong. Black-brown ; antennae rust-red ; variegated with yellowish-gray pubescence ; rostrum, a spot on thorax before scutellum ; scutellum, apex of elytra and pygidium white ; alternate interstices on elytra checkered with black and white, base of third interstice with a tubercle. L. 1¾ l. Rare.   *T. niveirostris*, Fab.

B. Transverse ridge on thorax curved, with convexity directed forward.

Oblong. Black-brown ; antennae and legs yellow-red, latter with black rings ; head, rostrum, disc of thorax covered with yellowish-gray hairs ; elytra unequally covered with grayish pubescence, with black and white markings, with a large, common dark spot on suture somewhat behind middle. Disc of thorax with two tubercles. L. 2 l. Rare.
    T. sepicola, Herbst.

Anthribus.

Oblong, convex. Black-brown ; very closely covered

with brown and gray hairlike scales; rostrum, forehead, a small spot somewhat before middle of each elytron, a broad band before their extreme apex and the under-side with white scales. Thorax in middle with three tubercles (placed in a transverse row) covered with black hairs; elytra with feeble rows of punctures, third interstice with three or four black tufts. Antennae in male with apex of all joints, besides apical part of eighth and basal part of ninth, in female only eighth joint closely covered with white scales. L. 3¼—4 l. Rare. *A. albinus*, Lin.

Brachytarsus.

A. Lateral elongations of transverse basal ridge on thorax strong and reaching apex.

Ovate. Black; elytra red, alternate interstices somewhat raised and checkered with spots of black and white hairs. Punctuation close; eyes prominent; posterior angles of thorax prominent, acute; elytra with punctured striae. L. 1½—1¾ l. Not very common.

B. scabrosus, Fab.

B. Lateral elongation of transverse basal ridge on thorax not very strong and scarcely reaching middle of sides.

Ovate. Black-brown, dull; with fine yellow-gray hairs, closer beneath than above; thorax with gray lines; elytra with rather deep punctured striae, sprinkled with gray, almost quadrangular spots; eyes prominent; posterior angles of thorax prominent, acute. L. 1—1¾ l. Not uncommon. *B. varius*, Fab.

Choragus.

Pitch-black; base of antennae yellow; base of femora, tibiae and tarsi yellow-brown. Thorax extremely finely punctured; elytra striate, deeply punctured. L. 1 l. Not common. *C. Sheppardi*, Kirb.

PHYTOPHAGA.

A. Head not hidden beneath thorax.

a. Head produced into a very short, flat rostrum.

Palpi threadlike, last joint nearly cylindrical or somewhat oval; antennae thickened toward apex, sawlike or comblike (in *Urodon* with three-jointed club); pygidium not covered. *Bruchidae.*

b. Head not produced into a rostrum.

I. Head constricted behind; thorax much narrower at base than elytra.

1. Apex of mandibles entire.

Labrum bilobed; antennae inserted before eyes.
Sagridae.

2. Apex of mandibles bifid or trifid.

A A. Eyes round.

Antennae long; inserted before eyes; first abdominal segment as long as the rest together; legs long.
Donaciidae.

B B. Eyes emarginate.

Antennae inserted within front inner margin of eyes.
Crioceridae.

II. Head not constricted behind; thorax not much narrower at base than elytra.

1. Posterior femora not thickened; insects without power of leaping.

A A. Head vertical.

a a. Antennae short, sawlike, or with larger apical joints.

Thorax double as broad as long, eyes round or oblong.
Clythridae.

b b. Antennae long, threadlike.

Elytra not entirely covering pygidium; eyes kidney-shaped. *Cryptocephalidae.*

B B. Head more or less prominent.

a a. Antennae not long, placed somewhat apart and slightly thickened toward apex.

Palpi short; eyes oblong; legs equal in size; tibiae without apical spines; body hemispherical or oval.
Chrysomelidae.

b b. Antennae long, placed closed together, not thickened toward apex.

Eyes entire; legs nearly uniform; body ovate, widest behind. *Galerucidae.*

2. Posterior femora thickened; insects with power of leaping.

Antennae inserted between eyes, usually close together; anterior coxae placed apart. *Halticidae.*

B. Head hidden beneath thorax.

Antennae short, slightly thickened toward apex; legs contractile, tibiae without spines, tarsi broad; body nearly circular or oval, slightly convex above, flat beneath.

Cassididae.

BRUCHIDAE.

A. Antennae ending in a distinct three-jointed club.

Head without neck behind; antennae inserted on sides of rostrum, under a feeble border, near eyes; pronotum rounded at sides and separated from sides of prothorax by a fine suture placed more or less low; anterior coxae small, globular, prosternum flat between them; tibiae not spined at apex. *Urodon,* Schoenh.

B. Antennae thickened toward apex, sawlike or comblike.

Head with neck behind; antennae inserted on upper part of rostrum, immediately before eyes; pronotum sharply bordered at sides; anterior coxae large, oval or transverse, prosternum arched between them; posterior coxae narrow, not covering first abdominal segment; tibiae spined at apex or having their inner angle prominent and acute. *Bruchus,* Lin.

Urodon.

Ovate. Black; antennae and legs red-yellow, hinder pairs of femora black at apex; closely covered with gray pubescence. L. 1 l. Very rare : Leicester.

U. rufipes, Fab.

Bruchus.

A. Antennae of male comblike.

Short ovate. Pitchy-red ; thorax with two united white tubercles at base ; scutellum white ; elytra with rather remote punctured striae, with obscure dark spots at base,

R 2

two irregular whitish transverse bands in middle, and a more distinct white spot near base of suture ; legs dull red, femora sometimes black at base. Antennae of male comblike, of female sawlike. L. 1¼—1½ l. Rare.

B. pectinicornis, Lin.

B. Antennae of male not comblike.

a. Thorax conical ; antennae entirely black.

I. Pubescence scarcely altering black ground colour.

Oblong-ovate. Black ; with very fine gray pubescence (scarcely altering the black colour). Second and third joints of antennae small, fourth only a little larger than third ; thorax finely and closely granulate ; elytra with very distinct punctured striae ; femora not toothed. L. ¾ l. Rather common. *B. cisti*, Fab.

II. Pubescence obscuring black ground colour.

Oblong ovate. Black, evenly covered with fine grayish-white pubescence. Thorax conical, closely and finely granulate ; elytra with distinct punctured striae ; antennae sawlike from fourth joint ; femora not toothed. L. ¾ l. Rare. *B. canus*, Germ.

b. Thorax transverse ; antennae with at least base reddish.

I. Sides of thorax toothed.

1. Legs partly yellow-red.

A A. Anterior femora entirely black.

Oblong-ovate. Black ; first four joints of antennae, anterior tibiae and tarsi, and often apex of intermediate tibiae yellow-red ; pubescence rather close, whitish-gray, placed in spots ; pygidium closely covered with white pubescence, with two large, ovate, black spots. Femora toothed. L. 2 l. Rare. *B. pisi*, Lin.

B B. Anterior femora wholly or partly yellow-red.

a a. Antennae with first four joints yellow-red.

A a. Pygidium with two more or less distinct dark spots.

A 1. Lateral teeth on thorax not strong.

Very similar to *B. pisi*, but with thorax longer, and more closely punctured ; tooth on femora smaller, obtuse ; black spots on pygidium often indistinct ; anterior femora entirely yellow-red. L. 1¾ l. Common. *B. rufimanus*, Boh.

B 1. Lateral teeth on thorax strong.

Ovate. Black ; first four joints of antennae and anterior legs yellow-red ; thorax with a spot in middle of base, elytra with base of suture and two bands of spots (near each other) closely covered with white hairs. Sides of thorax more deeply sinuate behind than in *B. rufimanus ;* posterior femora with a small, acute tooth. L. 1½ l. Rare. *B. affinis*, Froeh.

B b. Pygidium without dark spots.

Ovate. Black ; first four (rarely three) joints of antennae and anterior legs (except more or less of femora) yellow-red ; somewhat variegated with gray pubescence. Intermediate tibiae of male with a small tooth near apex ; posterior femora deeply emarginate beneath before apex. L. 1½ l. Rather common.

B. atomarius, Lin. (*seminarius*, Sharp's Cat.)

b b. Antennae entirely yellow-red.

Ovate, flat. Black ; mouth, antennae and front pairs of legs yellow-red ; pubescence gray, thorax with a spot in middle of base and several small scattered spots, the scutellum, elytra with base of suture, and a double, interrupted, transverse line on middle white, sides of abdomen with white points. Posterior femora distinctly toothed. L. 1⅓ —1½ l. Moderately common. *B. luteicornis*, Ill.

c c. Antennae with first five joints yellow-red.

Ovate. Black ; first five joints of antennae and front pairs of legs (except anterior femora about to middle and nearly all intermediate femora) yellow-red ; pubescence gray, thorax with two small spots on disc and one in middle of base, the scutellum, elytra with base of suture and many small spots whitish. Femora toothed. L. 1½ l. Rare. *B. nubilus*, Boh.

2. Legs entirely black.

Ovate. Black ; antennae with first four joints rust-red, second joint short ; thorax and elytra variegated with spots

of whitish pubescence. Femora toothed. L. 1¼ l. Rare.
B. viciae, Ol.

II. Sides of thorax not toothed.

1. Femora distinctly toothed.

A A. Elytra without spots.

Ovate. Black; mouth, first four joints of antennae and anterior legs (except base of femora) yellow-red; pubescence scanty, gray, evenly distributed. L. 1—1½ l. Scarce.
B. lathyri, Steph., (*loti*, Sharp's Cat.)

B. Loti, Payk. has a spot of close pubescence on scutellum and base of suture.

B B. Elytra with spots.

Ovate, rather flat. Black; antennae short, their first four or five joints and front pairs of legs (except intermediate femora) red-yellow; pubescence close, brown, diffusely spotted with white; thorax short, with two black spots on disc. L. 1⅓—1½ l. Rare. *B. lentis*, Boh.

2. Femora not toothed.

Ovate. Black; base of antennae red-brown; pubescence scanty, gray, uniform. Antennae almost as long as half body, gradually thickened toward apex. L. 1⅓—1¼ l. Rather common. *B. ater*, Marsh.

SAGRIDAE.

Antennae threadlike; eyes round, prominent, thorax much narrowed behind; elytra much broader than thorax; apex of tibiae with two large spines and a ring of bristles; tarsal claws split. *Orsodacna*, Latr.

Orsodacna.

A. Upper-side bare.

Elongate. Colour variable; usually antennae, thorax and legs reddish-yellow, elytra pale, breast and often abdomen black; sometimes with thorax and margins of elytra blackish; at other times body black, with antennae and legs, frequently also thorax and front of head reddish-yellow; elytra often black-blue. Thorax scantily punctured, elytra rather more strongly so. L. 2—3½ l. Not common. *O. cerasi*, Fab.

B. Upper-side hairy.

Head and breast black; thorax and abdomen reddish-yellow, former usually with a black spot; elytra and legs pale yellow-brown, suture of former often blackish. Thorax closely punctured. L. 2—2½ l. Scarce.

O. nigriceps, Latr.

The variety *O. humeralis,* Latr., has the upper-side black-blue, with a reddish spot at the shoulders, under-side black, base of antennae red-yellow and legs pitchy.

DONACIIDAE.

A. Third joint of tarsi bilobed, last joint not very long.

Thorax more or less square, usually with a little prominence at anterior angles; apex of elytra not produced; tarsal claws simple. *Donacia,* Fab.

B. Third joint of tarsi not bilobed, last joint very long.

Thorax as broad as long; outer apical angle of elytra produced into a point; tarsal claws simple.

Haemonia, Lac.

Donacia.

A. Elytra flat.

a. Posterior femora of male with two teeth; posterior tibiae with or without prominences.

I. Posterior tibiae of male with small prominences on inner-side.

1. Disc of thorax scarcely punctured.

Broad. Upper-side metallic-green, usually with some bluish reflection; under-side covered with silvery hairs; base of antennae and legs (except their upper-side) reddish. Thorax with a prominence on each side in front, the sides very finely shagreened, disc scarcely punctured; elytra with strong punctured striae, inner interstices smooth, outer ones finely wrinkled transversely. Posterior femora of male with two teeth (the front one pointed, the hinder one broader and stronger), those of female with one small tooth. L. 4—5 l. Moderately common.

D. crassipes, Fab.

2. Disc of thorax diffusely punctured, somewhat wrinkled.

Broad, rather convex. Bronze, with coppery reflection, under-side with silvery pubescence; base of antennae and legs reddish, club of femora bronze. Thorax as long as broad, disc diffusely punctured, somewhat wrinkled ; elytra with deep punctured striae, interstices transversely wrinkled, apex truncate. Posterior femora of male with two teeth placed transversely, those of female with one small tooth. L. 3½—4 1. Moderately common. *D. bidens*, Ol.

3. Disc of thorax closely wrinkled.

Upper-side golden-green or coppery; under-side covered with silvery hairs ; base of antennae and legs (except apex of femora) reddish. Thorax broader than long, closely punctured in wrinkles ; apex of elytra obliquely truncate ; first abdominal segment of male somewhat impressed in middle, with two blunt elevations. Posterior femora of both sexes with two teeth placed transversely, the inner one being very small in female. L. 3½—4 1. Not uncommon. *D. dentata*, Hoppe.

II. Posterior tibiae of male without prominences on inner-side.

Upper-side green ; under-side with gray pubescence. Thorax quadrangular, with angles scarcely prominent and a blunt elevation on each side in front, surface finely and closely scratched, central furrow fine ; elytra moderately narrowed toward apex, which is cut off straight, with punctured striae and some impressions, interstices wrinkled. Posterior femora with two teeth in both sexes, front one indistinct in female. L. 3½—4½ 1. Not common. *D. sparganii*, Ahr.

b. Posterior femora of male with only one tooth ; posterior tibiae without prominence.

I. Disc of elytra even, or with only one impression rather before middle near suture.

Bronze ; elytra golden-green, each with a broad longitudinal streak (often indistinct) of purple ; under-side and legs with yellow pubescence. Thorax as long as broad, punctured in wrinkles, with central furrow, anterior angles bluntly prominent ; elytra much narrowed behind, apex cut off straight, with fine punctured striae, interstices transversely wrinkled. Posterior femora in male more strongly toothed than in female. L. 3—4 l. Common.

D. dentipes, Fab.

II. Disc of elytra with two or more distinct impressions.

1. Thorax with a deep central furrow.

A A. Elytra unicolorous or rather darker on disc. '

Upper-side green, usually with golden reflection ; under-side with golden-yellow pubescence ; head and thorax rarely blue. Thorax, elytra and posterior femora as in *D. lemnae*. L. 4—4½ l. Rather common.

D. sagittariae, Fab.

B B. Elytra brown-bronze, with a dark purple longitudinal streak (often indistinct) near side margin and a spot at base.

Head and thorax brown-bronze ; under-side with yellow-ish-gray pubescence. Thorax quadrangular, coarsely punctured in wrinkles, anterior angles not very prominent, the elevation on each side indistinct ; elytra deeply punctured in rows, interstices finely transversely wrinkled, each elytron with two distinct, shallow impressions. Posterior femora with a small tooth before apex, often indistinct in female. L. 4—4½ l. Common. *D. lemnae*, Fab.

2. Thorax with a fine central furrow.

A A. Forehead without prominence ; posterior femora distinctly toothed.

a a. Anterior angles of thorax prominent.

Upper-side brown-bronze, not very shiny; under-side with golden-yellow pubescence. Thorax tolerably quadrangular, distinctly narrowed behind, punctured in wrinkles ; elytra with distinct impressions, with punctured striae, interstices very closely wrinkled transversely. L. 4—4½ l. Rare. *D. obscura*, Gyll.

b b. Anterior angles of thorax not prominent.

Upper-side bronze ; under-side with yellow pubescence. Thorax as long as broad, narrowed behind, punctured in wrinkles ; elytra with punctured striae, interstices very finely wrinkled transversely. L. 4 l. Rather common.

D. thalassina, Germ.

B B. Forehead with a strong, blunt prominence
on each side ; posterior femora indistinctly
toothed.

Elongate.   Upper-side coppery ; under-side with yellow
pubescence.   Thorax tolerably quadrangular, with scarcely
any prominences in front, punctured in wrinkles ; elytra
with four distinct impressions, interstices of punctured
striae finely wrinkled transversely.   L. 3 l.   Rather common.   *D. impressa*, Payk.

c. Posterior femora without any tooth in both sexes,
posterior tibiae without prominences.

I. Upper-side not hairy.

1. Apex of elytra truncate.

A A. Elytra more than double as long as
together broad.

a a. Elytra unicolorous or nearly so.

Elongate, rather flat.   Dull bronze-green or purple ;
under-side with silvery pubescence ; antennae black, base
of each joint lighter ; femora bronze, their base and the
curved tibiae dull red-yellow.   Thorax quadrate, closely
punctured, apical margin slightly rounded ; elytra with
four depressions, interstices of punctured striae transversely wrinkled, apex truncate, outer angle distinct.
L. 4—5 l.   Common.   *D. linearis*, Hoppe.

b b. Elytra with a distinct purple streak
near suture on each.

Upper-side metallic-green, with a purple streak on each
elytron, rarely entirely coppery or purple ; under-side with
silver-gray pubescence.   Thorax very closely punctured,
with a prominence on each side in front, anterior angles
blunter than in *D. semicuprea,* apical margin distinctly
rounded ; elytra with some feeble depressions, punctured
deeply in rows, interstices finely wrinkled transversely,
apex cut off straight or slightly emarginate, outer angle
distinct.   L. 3½—4½ l.   Rather common.

D. typhae, Brahm.

B B. Elytra less than twice as long as together
broad.

Upper-side green or bronze-green, elytra coppery on

disc; under-side with silver-gray pubescence. Thorax very closely punctured, with a prominence on each side in front and a central furrow or depression on basal part of disc; elytra with rows of deep punctures, interstices finely wrinkled transversely, outer apical angle rounded. Posterior femora not reaching apex of elytra. L. $3\frac{1}{2}$—4 l. Common. *D. semicuprea*, Panz.

2. Elytra separately rounded at apex.

Upper-side golden-green; under-side with close silvery pubescence, antennae and legs reddish. Thorax almost longer than broad, finely wrinkled transversely, with central furrow, with a prominence on each side in front, middle of sides slightly sinuate; elytra with rows of deep punctures and very fine wrinkles. Posterior femora reaching apex of elytra. L. $4\frac{1}{2}$ l. Rather common. *D. menyanthidis*, Fab.

II. Upper-side closely pubescent.

Bronze; pubescence on upper-side very fine, gray, on under-side close, silvery; base of femora and often of antennae red. Thorax as long as broad, with a blunt prominence on each side in front, slightly narrowed behind; elytra with very fine punctured striae and flat interstices. L. $3\frac{1}{2}$—$4\frac{1}{2}$ l. Rather common.

D. hydrochaeridis, Fab.

B. Elytra more or less convex.

a. Posterior femora with a strong, triangular tooth in both sexes; each elytron with two small depressions; legs concolorous with body.

I. Third joint of antennae distinctly longer than second; elytra narrowed from shoulders.

Upper-side purple, golden-green, blue or coppery; under-side with golden-yellow pubescence. Thorax finely wrinkled, with a deep central furrow, with a large prominence on each side in front, anterior angles scarcely prominent; elytra with punctured striae, interstices finely wrinkled transversely, apex rounded. L. 3—$3\frac{1}{2}$ l. Common.

D. sericea, Lin.

II. Second and third joints of antennae about equal in size; elytra parallel-sided until near apex. .

Similar to *D. sericea* but broader, antennae and legs

shorter and thicker; fourth joint of antennae only a little larger than third; anterior angles of thorax scarcely perceptible, the prominence gradually absorbed by angle; elytra more coarsely punctured. L. 3—3½ l. Rather common.   *D. comari*, Suffr.

b. Posterior femora of male with a strong, of female with a feeble tooth ; elytra without depressions ; legs reddish.

I. Length 4½—5½ lines.

Upper-side black, with bluish reflection, thorax more or less violet; under-side with gray pubescence; abdomen, antennae and legs red-brown. Thorax with a large, blunt prominence on each side and obtuse anterior angles ; elytra with punctured striae, the punctures being much crowded and somewhat indistinct when viewed obliquely, the interstices being strongly wrinkled. Rather common.

        D. nigra, Fab.

II. Length 3—3½ lines.

Male with thorax narrowed behind and sinuate at sides, moderately finely and scantily punctured, with obtuse, raised anterior angles ; elytra purplish-black, with coarsely punctured striae, interstices transversely wrinkled and punctured at shoulders and scutellum. Female with thorax much narrowed behind, sinuate at sides, finely and rather closely punctured, anterior angles obtuse, raised ; elytra bronze, with coarsely punctured striae, interstices coarsely wrinkled transversely. Antennae in both sexes red. Rather common.   *D. affinis*, Kunze.

Haemonia.

A. Spine at apex of elytra moderately long; length 2¾—3¼ lines.

Yellow; head, breast and antennae black and covered with thick whitish pubescence; disc of thorax with two short black streaks; elytra pale yellow, with black punctured striae (placed in pairs). Second and third joints of antennae short ; second tarsal joint much longer than first and third. Rare.   *H. equiseti*, Fab.

B. Spine at apex of elytra rather short; length 2¼—2½ lines.

Similar to *H. equiseti*, but elytra with apical spine

blunter and shorter, the striae blacker and the punctures finer. Not uncommon. *H. Curtisi*, Lac.

CRIOCERIDAE.

A. Tarsal claws toothed.

Head and thorax broader than long, latter with a blunt prominence on each side. *Zeugophora*, Kunze.

B. Tarsal claws simple.

a. Tarsal claws soldered together at base.

Forehead deeply furrowed ; thorax dilated in front, narrowed behind middle or at base ; intermediate coxae placed apart. *Lema*, Fab.

b. Tarsal claws free throughout.

Similar in other respects to *Lema*. *Crioceris*, Lac.

Zeugophora.

A. Elytra blackish.

a. Head entirely red-yellow.

Black, shiny, with fine gray pubescence ; head, base of antennae, thorax, prosternum and legs red-yellow. Thorax and elytra strongly punctured, former with a smooth central line, its lateral projection forming a feeble angle with anterior margin nearly on a level with inner margin of eye. L. $1-1\frac{1}{4}$ l. Common. *Z. subspinosa*, Fab.

b. Head red-yellow in front and beneath, with forehead and vertex black.

Black, shiny ; front and under-side of head, base of antennae, thorax, prosternum and legs reddish-yellow. Thorax and elytra coarsely and deeply punctured, lateral prominences on former pointed and usually a central line smooth, latter often with yellow shoulders. L. $1\frac{1}{2}$ l. Rare. *Z. flavicollis*, Marsh.

B. Elytra red-yellow.

Head, thorax, elytra, antennae and legs red-yellow ; eyes, mesosternum, metasternum and abdomen black ; apex of mandibles slightly pitchy. Head more closely and deeply punctured than in *Z. subspinosa*, less contracted behind eyes, latter less prominent ; thorax broader, lateral projection more prominent, more abruptly produced and continued with a slight curve until it meets the anterior margin, the junction forming a distinct angular process on a level with outer margin of eye. Rare. *Z. Turneri,* Power.

Lema.

A. Thorax blue; legs black or black-green.

a. Thorax narrowest near middle.

Convex. Blue; antennae black; under-side and legs finely pubescent. Thorax nearly heart-shaped, sides of constriction closely striate, disc with two rows of punctures; elytra with not very strong punctured striae; posterior tibiae very slightly curved. L. $1\frac{3}{4}$—2 l. Not uncommon.

L. puncticollis, Curt.

b. Thorax narrowest near base.

I. Sides of thorax not punctured behind, sides of constriction striate.

Blue or blue-green, antennae and legs black. Elytra with rows of strong and deep punctures. Upper-side rarely dull black. L. $1\frac{3}{4}$—2 l. Common.

L. cyanella, Fab.

II. Sides of thorax punctured behind, sides of constriction punctured.

Similar to *L. cyanella,* but elytra more finely punctured, interstices transversely wrinkled, with a few punctures. L. 2—$2\frac{1}{4}$ l. Rare. *L. Erichsoni,* Suffr.

B. Thorax and legs (except tarsi) yellow-red.

Head, antennae and tarsi black; elytra blue or blue-green, with rows of large, deep punctures. L. $1\frac{1}{2}$—$1\frac{3}{4}$ l. Common. *L. melanopa,* Lin.

Crioceris.

A. Elytra entirely red.

Black; thorax and elytra bright red (fading after death). Thorax with scattered punctures on central line; elytra with fine punctured striae, the punctures stronger at base and apex than on disc. L. $3\frac{1}{4}$—$3\frac{1}{2}$ l. Scarce.

C. merdigera, Fab.

B. Elytra red-yellow, with six small black spots on each.

Head, thorax and legs yellowish-red; scutellum, antennae, eyes, breast, knees and tarsi black. L. 2½ l. Rare. *C. 12-punctata*, Lin.

C. Elytra blue-green, with red sides and three spots on each (often united) whitish-yellow.

Head, antennae and under-side blue-green; thorax red, with middle dark; legs dark, base of tibiae sometimes lighter. Thorax as long as broad, scarcely dilated at sides, not constricted behind, disc with scattered punctures. L. 2½ l. Common. *C. asparagi*, Lin.

CLYTHRIDAE.

A. Antennae sawlike; body oblong.

Clypeus usually emarginate; thorax almost more than twice as broad as long, usually somewhat produced in middle of base. *Clythra*, Laich.

B. Antennae with five larger apical joints; body oval, very convex.

Mouth for most part covered by prominent prosternum; thorax double as broad as long, fitting closely to elytra, posterior angles right angles. *Lamprosoma*, Kirby.

Clythra.

A. Elytra unicolorous.

Moderately elongate. Metallic-green or blue-green, with short white pubescence beneath; elytra, labrum and palpi yellow; antennae violet, with yellow base. Head closely punctured in wrinkles, forehead scarcely impressed; thorax rather closely and distinctly punctured, posterior angles produced and raised; elytra closely punctured; emargination of clypeus with a distinct tooth; fourth joint of antennae cylindrical, fifth sawlike; anterior tibiae curved. L. 3—3½ l. Moderately common. *C. tridentata*, Lin.

B. Elytra yellow or yellow-red, marked with black.

Base of thorax completely rounded; both fourth and fifth joints of antennae sawlike; first tarsal joint much shorter than the next two together.

a. Sides of thorax broadly bordered and scarcely raised, disc punctured.

Black, not very shiny; elytra shiny yellow-red, each with a black spot at shoulder and a larger transverse one (often divided into two) behind middle; head and under-side with gray pubescence. Thorax very short, not very convex, closely and unequally punctured. L. 3—5 l. Moderately common.   *C. quadripunctata*, Lin.

b. Sides of thorax narrowly bordered and raised, disc impunctate.

Black, shiny; elytra yellow, each with a black spot at shoulder and a broad black band (interrupted only narrowly by suture and nearly reaching side margin) some-what behind middle; under-side and head with gray pubescence. Thorax short, convex, impunctate. L. 3—5 l. Rare.   *C. laeviuscula*, Ratz.

Lamprosoma.

Black, with brownish reflection. Upper-side very finely confusedly punctured, elytra also with rows of punctures. L. 1—1¼ l. Not uncommon.   *L. concolor*, Sturm.

CRYPTOCEPHALIDAE.

Head hidden; thorax more or less cylindrical or globular, apex slightly emarginate, base somewhat produced in middle; scutellum large; elytra short, cylindrical, sepa-rately rounded at apex.   *Cryptocephalus*, Geoffr.

Cryptocephalus.

A. Elytra confusedly punctured, or with irregular rows of punctures here and there only.

a. Elytra red or yellow, with or without black markings.

I. Femora entirely black.

1. Elytra brick-red, with five black spots on each.

Black; elytra as above; base of antennae brown. Thorax punctured; elytra with scattered punctures. L. 2½—2⅔ l. Rare.   *C. imperialis*, Fab.

2. Elytra red, usually with black shoulders.

Black, shiny; elytra as above; base of antennae and two lines on forehead yellow; female also with red thorax. Thorax very finely, elytra more coarsely punctured, latter here and there in rows. L. 2½—3 l. Not common.

C. coryli, Lin.

II. Femora with white spot at apex.

Black; a spot on clypeus, anterior and side margins of thorax yellow, disc of latter with a whitish-yellow central line, abbreviated behind in male, split in female; elytra red, with variable black markings ; base of antennae brown. Elytra coarsely punctured. L. 2¼ l. Scarce.

C. sexpunctatus, Lin.

b. Elytra green-bronze-green, or purplish.

I. Thorax distinctly punctured ; antennae entirely black.

1. Sides of thorax somewhat sinuate before posterior angles, rounded in middle.

Golden-green or purplish, frosted above. Elytra punctured in wrinkles. L. 2¾ l. Common.

C. aureolus, Suffr.

2. Sides of thorax only slightly sinuate before anterior angles, thence straight.

Golden-green, bronze-green or purplish, frosted above. Elytra punctured in wrinkles. L. 2—2¾ l. Rather common. *C. hypochaeridis*, Lin.

II. Thorax impunctate ; antennae yellow at base.

Green or golden-green, smooth and shiny ; under-side black ; under-side of head, a frontal spot, and base of antennae yellow ; male with the anterior legs and all tibiae, female with all the legs yellow. Thorax double as broad as long, generally yellow in front and at sides in male ; elytra very coarsely, scantily punctured. L. 1½—2 l. Rather common. *C. nitidulus*, Gyll.

B. Elytra with punctured striae.

a. Elytra black, with a yellow spot at apex and another at side margin.

Black, shiny ; base of antennae, a frontal spot, posterior angles (sometimes anterior margin and sides) of thorax, two spots on elytra as above, greater part of anterior femora and the tibiae (posterior pair sometimes excepted) yellow. Thorax very finely punctured; elytra with deep punctured striae. L. 1½—2 l. Rather common.

C. moraei, Lin.

b. Elytra red or yellow, with or without black spots, or entirely blue or black.

I. Thorax yellow, spotted with black.

Black; base of antennae, a frontal spot, thorax (except spots) and legs yellow; elytra varying from yellow, with five black spots on each, to entirely black. L. 1½—2 l. Rare.   *C. decempunctatus,* Lin.

II. Thorax black, with or without yellow anterior margin, or entirely yellow.

1. Elytra entirely blue.

A A. Front pairs of legs yellow.

Elongate.   Black; elytra blue; base of antennae, a forked frontal spot, mouth, anterior margin of thorax and front pairs of legs yellow. Thorax coarsely and deeply punctured, more scantily on disc; elytra with coarsely punctured striae. L. ¾—1¼ l. Rare. *C. punctiger,* Payk.

B B. Legs blue.

Dull violet; under-side of head, base of antennae and coxae yellow; male also with anterior margin of thorax yellow. Thorax transversely impressed behind; interstices on elytra smooth. L. 1¾ l. Not common.
                                        C. fulcratus, Germ.

2. Elytra red or yellow, with or without black markings.

A A. Legs black; length 2½ lines.

a a. Elytra red, with a longitudinal black line on disc of each.

Black, shiny; elytra red, with outer margin, suture and a longitudinal line on disc black; base of antennae pitchy. Not uncommon.   *C. lineola,* Fab.

b b. Elytra black, with a waved red spot at apex of each.

Black, shiny; elytra with a waved red spot at apex of each; base of antennae pitchy. Scarce.
                                      C. bipustulatus, Fab.

B B. Legs reddish-yellow; length ¾—1¼ lines.

a a. Thorax black, with anterior margin and sides yellow.

Black; thorax as above; elytra pale yellow, with suture and a broad longitudinal streak on each black; male with greater part of forehead, female with its apex only and two spots on vertex yellow. Thorax closely covered with very fine wrinkles; elytra with deep punctured striae. L. 1 l. Common. *C. bilineatus*, Lin.

b b. Thorax entirely yellow.

A a. Elytra with punctured striae throughout.

Head, thorax and legs red-yellow; elytra yellow, usually with suture and rarely shoulders black; breast and abdomen black, with a yellow-brown spot beneath shoulders. Thorax impunctate. L. 1 l. Common.

C. minutus, Fab.

B b. Punctured striae on elytra almost effaced from middle.

Brown-yellow; breast and abdomen black; elytra either unicolorous or marked with black. Thorax impunctate. L. ¾—1¼ l. Common. *C. pusillus*, Fab.

3. Elytra entirely black.

A A. Thorax entirely black.

a a. Antennae black at apex.

A a. At least inner striae on elytra effaced toward apex.

Black; front of head, base of antennae and the legs reddish-yellow. Thorax smooth. L. 1—1¼ l. Common. *C. labiatus*, Lin.

B b. Striae on elytra regular throughout.

Black; base of antennae and the legs yellow, with posterior femora darker; male with under-side of head and two oblique lines on forehead yellowish; female with mouth only pitch-brown. Thorax with fine scratches. L. 1—1¼ l. Rare. *C. Wasastjernae*, Gyll.

b b. Antennae entirely yellow.

Black; antennae, under-side of head and legs yellow; male with front of head yellow; female with only mouth brownish. Thorax impunctate; striae on elytra finer from middle backward. L. 1¼—1½ l. Rare .*C. querceti*, Suffr.

B B. Thorax black, with anterior margin yellow.

s 2

Black; base of antennae, under-side of head, a large frontal spot, anterior margin of thorax and legs yellow, femora brownish; scutellum of male yellow. Thorax smooth; elytra with close punctured striae. L. 1—1¼ l. Scarce.   *C. frontalis*, Marsh.

CHRYSOMELIDAE.

A. All tarsal joints equal in breadth ; insects apterous.

Anterior coxae placed only slightly, hinder pairs far apart ; tibiae without grooves for reception of tarsi.
                                                   Timarcha, Redt.

B. Second tarsal joint narrower than first ; insects winged.

a. Body not elongate.

I. Antennae threadlike or gradually thickened.

1. Tarsal claws simple.

A A. Tibiae without furrows on outer-side, or if furrowed, thorax almost as broad at base as elytra.

Antennae reaching beyond base of thorax, sometimes threadlike ; mesosternum short, excised in front ; anterior coxae placed apart.   *Chrysomela*, Lin.

B B. Posterior tibiae furrowed on outer-side ; thorax narrower at base than elytra.

Antennae scarcely reaching beyond base of thorax, considerably thickened toward apex ; mesosternum very short, excised in front ; anterior coxae placed apart.
                                                       Lina, Redt.

2. Tarsal claws toothed at base.

Antennae as long as half the body, gradually thickened ; mesosternum very short ; anterior coxae almost contiguous ; tibiae more or less furrowed on outer-side, toothed on outer-side of apex.   *Gonioctena*, Redt.

II. Antennae more or less distinctly clubbed.

1. Posterior tibiae toothed on outer-side of apex.

Antennae scarcely reaching middle of body, the last six joints forming a sort of elongate club ; mesosternum narrow, triangular and pointed ; anterior coxae placed very close together.   *Gastrophysa*, Chevr.

2. Posterior tibiae not toothed at apex.

A A. Reflexed margin of elytra strongly furrowed ; first ventral segment of abdomen a
little longer than second and much shorter than
second and third together.

Antennae short, not one-third as long as body, last five
joints forming an almost distinct club; mesosternum very
broad and very short; elytra confusedly punctured.

Plagiodera, Redt.

B B. Reflexed margin of elytra even; first ventral segment of abdomen nearly as long as
next three together.

Antennae rather short, not reaching middle of body, last
five joints forming an elongate club; mesosternum very
short, broadly sinuate in front; elytra punctured in rows.

Phaedon, Latr.

b. Body elongate.

I. Tarsal claws toothed at base.

Antennae a little shorter than half the body, slender at
base, gradually thickened at apex ; mesosternum rather
broad and long ; elytra punctured in rows.

Phratora, Redt.

II. Tarsal claws simple.

Antennae scarcely reaching beyond base of thorax, last
five joints forming a distinct club ; mesosternum as broad
as, or broader than prosternum, not excised at base ; head
somewhat produced in front, last joint of palpi long.

Prasocuris, Latr.

Timarcha.

A. Thorax broadest before middle, much more narrowed
behind than before.

Black, dull ; under-side and legs dark-blue or blue-green.
Thorax very closely and finely punctured ; elytra as finely
but more scantily punctured. L. 5—6 l. Common.

T. laevigata, Lin.

B. Thorax broadest about in middle, not much more narrowed behind than before.

Black, almost lustreless, usually with violet reflection ;
legs dark-blue or blue-green. Punctuation of thorax fine

and close, of elytra tolerably deep and unequal.   L. 4—5 l.
Common.   *T. coriaria*, Fab.

Chrysomela.

A. Elytra confusedly punctured, with or without irregular
double rows of larger punctures.

  a. Upper-side unicolorous bronze or dark blue.

   I. Antennae slender; sides of thorax raised.

    1. Length 4—4½ lines.

Short oval.   Bronze or greenish-bronze; mouth, antennae and under-side rust-yellow.   Side border of thorax
narrowly separate and raised; elytra irregularly coarsely
punctured.   Common.   *C. Banksi*, Fab.

    2. Length 2½—3 lines.

Oval.   Yellow-brown, shiny.   Side border of thorax
separated by a broad impression and raised; punctuation
fine, elytra with irregular double rows.   Common.
                                        C. staphylaea, Lin.

   II. Antennae short; sides of thorax not raised.

    1. Elytra simply punctured.

     A A. Elytra coarsely punctured; length 2¼—2¾
         lines.

Short oval.   Dark blue, violet, dark green or bronze.
Sides of thorax shallowly impressed before posterior angles.
Rather common.   *C. varians*, Fab.

     B B. Elytra rather finely punctured; length
         3—4 lines.
Oval.   Blue; palpi, antennae and tarsi rust-yellow.
Not very common.   *C. goettingensis*, Lin.

    2. Elytra finely punctured, with an admixture of
        larger punctures, placed in irregular double rows.

Hemispherical.   Black-blue; base of antennae reddish.
Thorax closely and finely punctured.   L. 2¾—3¼ l.   Common.   *C. haemoptera*, Lin.

  b. Upper-side dark blue or brownish, with red margin
      to elytra.

   I. Elytra confusedly punctured.

    1. Punctuation of elytra deep, confluent.

Broad elliptic. Deep black-blue; sides of elytra red. Thorax parallel-sided behind middle, shortly rounded in front, side border separated by a wrinkled impression formed of coarse punctures ; elytra coarsely and irregularly closely punctured. L. 3½—4½ l. Not very common.

C. sanguinolenta, Lin.

2. Punctuation of elytra not deep and not confluent.

Ovate. Deep bluish-black ; sides of elytra broadly blood-red ; antennae black, with basal joints and the legs bluish. Disc of thorax finely punctured, its sides and the elytra coarsely and closely punctured, the punctures not confluent. L. 3½—4 l. Not uncommon.

C. distinguenda, Steph.

II. Elytra finely punctured, with irregular double rows of larger punctures.

Narrow elliptic. Brownish-bronze ; base of antennae and a narrow border to elytra yellow-red. Thorax narrowed in front, side border separate before posterior angles ; elytra with alternate interstices slightly raised. L. 2¾—3½ l. Not common.

C. marginata, Lin.

c. Upper-side green, with or without elytra more or less red or purplish. •

I. Side margin of thorax not thickened.

1. Length 3½—5 lines.

A A. Thorax narrowed in front; upper-side unicolorous.

Oblong-elliptic. Golden-green. Punctuation of thorax close and coarse, of elytra scattered, with shiny, scratched interstices. Male with an oblong depression at base of last abdominal segment. L. 3½—3¾ l. Common.

C. menthrasti, Suffr.

B B. Thorax not narrowed in front; thorax, suture of elytra and a badly defined longitudinal band on each of latter darker than rest of upper-side.

Oblong-elliptic, convex. Golden-green ; parts mentioned above darker. Punctuation of thorax moderately strong on disc, coarse at sides, of elytra coarsely wrinkled, with scratched interstices. L. 3¾—5 l. Rather common.

C. graminis, Lin.

2. Length 2¼—3 lines.

Oblong elliptic, flatly arched.   Golden-green ; suture of elytra and an abbreviated longitudinal band on each darker. Thorax parallel-sided, moderately strongly punctured (coarsely in the depressed posterior angles only) ; elytra evenly coarsely punctured, with more finely punctured and scratched interstices.   Common.   *C. fastuosa*, Lin.

II. Side margin of thorax thickened.

1. Elytra striped with purple, green, gold and copper.

Oblong elliptic.   Upper-side metallic-green or red ; base of antennae brownish ; thorax with three darker longitudinal lines ; elytra as above.   Thorax punctured, side border separated by a broad, interrupted impression ; elytra finely punctured.   L. 3—4 l.   Scarce.   *C. cerealis*, Lin.

2. Elytra unicolorous red-brown.

Oblong-elliptic.   Golden-green ; base of antennae red-yellow ; elytra red-brown, metallic.   Thorax punctured, side margin dilated in a curve in front, separated behind by a deep impression ; elytra coarsely punctured, interstices with isolated, fine punctures.   L. 2¼—3½ l.   Common.
                                                    C. polita, Lin.

B. Elytra with regular rows of punctures.

a. Elytra roughened, with simple rows of punctures, interstices with fine, scattered punctures.

Elliptic, much narrowed at ends.   Olive-green, shiny ; base of antennae reddish.   Lateral impressions of thorax deep behind ; interstices on elytra equal in breadth, finely and scantily punctured.   L. 3—3½ l.   Rather common.
                                                    C. lamina, Fab.

b. Each elytron with five double rows of coarse punctures, the first becoming simple behind middle and ending in a furrow, also with scattered fine punctures.

I. Body narrow elliptic.

Brassy, blue or blackish-bronze ; base of antennae brownish.   Punctuation of thorax effaced in front ; double rows of punctures on elytra coarse.   L. 2½—3 l.   Common.
                                                    C. hyperici, Forst.

II. Body broad elliptic.

Somewhat convex. Blue; base of antennae yellow-brown. Thorax finely punctured, sides separated by a fold behind. L. 2¾—3 l. Not uncommon.

C. didymata, Scrib.

Lina.

A. Thorax without separate side border.

Metallic-blue or green, rarely black; base of antennae and margin of apex of abdomen reddish-yellow. Thorax punctured finely and scantily on disc, more coarsely at sides. L. 2¾—4 l. Moderately common. *L. aenea*, Fab.

B. Thorax with separate side border.

a. Side border of thorax feebly separate, rounded behind.

Black blue; elytra red, with a small black spot at apex. Thorax coarsely punctured. L. 4½—6 l. Common.

L. populi, Lin.

b. Side border of thorax strongly separate, sinuate behind, with prominent posterior angles.

Blue-green; elytra red. Thorax coarsely punctured. L. 3½—4½ l. Rather common. *L. longicollis*, Suffr.

Gonioctena.

A. Body broad and flatly arched, interstices on elytra closely punctured; all tibiae distinctly spined.

a. Head (except vertex) and legs red.

Broad elliptic. Yellow-red; vertex of head, two spots at base of thorax, scutellum, five spots on each elytron and under-side black. Elytra finely and closely punctured, with distinct rows of larger punctures. L. 2½—3¼ l. Common. *G. rufipes*, Gyll.

b. Head and at least part of legs black.

I. Legs entirely black.

Broad elliptic. Black; thorax and elytra red, often spotted with black or even entirely black; base of antennae and apex of abdomen yellow-red. Elytra closely and distinctly punctured, with rows of larger punctures, almost effaced behind. L. 2½—3½ l. Common.

G. viminalis, Lin.

II. Femora and tarsi black, tibiae red.

Oblong-elliptic. Black; anterior angles of thorax and the elytra sometimes red, latter spotted with black; base of antennae, tibiae and apex of abdomen red. Thorax narrowed in front, anterior angles acute; elytra with rows of punctures and rough interstices. L. 2½—3 l. Rare.

G. affinis, Schoenh.

B. Body convex, oval; interstices on elytra finely and closely punctured; anterior tibiae inconspicuously spined.

Upper-side and legs reddish-yellow; a bilobed frontal spot, a band on each elytron and the under-side black; whole insect sometimes yellow. Elytra with coarse rows of punctures. L. 1¾—2 l. Common.   G. litura, Fab.

C. Body somewhat cylindrical; interstices on elytra finely and scantily punctured; anterior tibiae without spine.

Reddish-yellow, more or less marked with black, or even entirely black. Elytra with coarse rows of punctures; sutural apical angle distinctly produced; last abdominal segment simple; hinder pairs of tibiae equally strongly spined. L. 2½—3 l. Not uncommon.   G. pallida, Lin.

Gastrophysa.

A. Thorax red.

Blue-green; thorax, base of antennae, femora, tibiae and apex of abdomen red. L. 1¾—2¼ l. Common.

G. polygoni, Lin.

B. Thorax green.

Golden-green; club of antennae and mouth parts blackish. Posterior angles of thorax bluntly rounded. L. 2—2½ l. Common.   G. raphani, Fab.

Plagiodera.

Almost circular, very flatly arched. Black, upper-side blue, green, violet or coppery; base of antennae red-brown. Elytra confusedly punctured, with a furrow on shoulder. L. 1½—2 l. Scarce.   P. armoraciae, Lin.

Phaedon.

A. Thorax distinctly punctured at base only.

Short ovate. Brassy-black, punctulated; antennae pitch-black, basal joint pitchy-red. Head with a smooth impressed line; thorax very smooth, disc flattened, sides somewhat swollen ; elytra with punctured striae, interstices minutely punctured. L. 1½ l. Common.

P. *tumidulum*, Kirby.

B. Thorax distinctly punctured throughout.

a. Thorax evenly punctured ; transverse ridge between intermediate legs straight.

Oblong oval, flatly arched. Blue or blue-green, apex of abdomen red. Elytra with punctured striae, interstices distinctly punctured, outermost row consisting of isolated punctures only from middle backward, shoulders with a deep impression ; anterior margin of mesosternum triangularly, and broadly sharply excised. L. 1½—1¾ l. Common.

P. *betulae*, Lin.

b. Thorax more closely punctured at sides than on disc ; transverse ridge between intermediate legs curved.

I. Interstices on elytra very finely wrinkled ; anterior margin of mesosternum triangularly, sharply and broadly excised ; body flatly arched.

Oblong oval. Blue or green ; base of antennae and tarsi brownish. Elytra with fine punctured striae, the outermost row formed of united punctures, shoulders with a feeble impression. L. 1¼—1½ l. Moderately common.

P. *cochleariae*, Fab.

II. Interstices on elytra coarsely and closely punctured ; anterior margin of mesosternum excised in a deep curve ; body convex.

Oval, somewhat compressed laterally. Blue or green, shiny. Elytra with punctured striae and a feeble humeral impression ; ridge between intermediate coxae almost reaching anterior margin. L. 1¼—1½ l. Not uncommon.

P. *concinnum*, Steph.

Phratora.

A. Six inner striae on elytra irregular.

Metallic-blue-green, greenish or violet ; base of antennae and margin of abdomen reddish. Elytra with fine, rather wavy rows of punctures on disc and a longitudinal wrinkle behind shoulder. Male with posterior tibiae curved and first tarsal joint broader than third. L. 2¼—2½ l. Common.

P. *vulgatissima*, Lin.

B. Six inner striae on elytra regular.

a. Forehead broadly hollowed out.

Oblong. Bright blue, shiny; under-side of base of antennae and sides of anus (in male narrowly, in female broadly) red-yellow; tarsi brown. Forehead broadly hollowed out in middle. Differing from *P. vulgatissima* in having sides of thorax somewhat rounded before middle, posterior tibiae of male not curved, and first tarsal joint of male only slightly dilated; more oblong than *P. vitellinae*, with antennae longer and stouter, thorax less transverse, with sides rounded before middle, first tarsal joint of male less dilated and much narrower than third. L. 1¾ l. Not uncommon. *P. cavifrons*, Th.

b. Forehead not hollowed out.

Bronze or greenish; base of antennae and margin of abdomen reddish. Elytra with regular rows of coarse punctures on disc. First tarsal joint of male a little narrower than third. L. 1¾—2 l. Common.

P. vitellinae, Lin.

Prasocuris.

A. Thorax broader than long.

a. Thorax dilated before middle, unicolorous dark-green.

Short ovate, convex, somewhat compressed laterally. Dark green, with a broad yellow-red margin to elytra (sometimes absent). Thorax coarsely and closely punctured; elytra with rows of coarse punctures, the outermost broken into a series of connected punctures. L. 1¼—1½ l. Rather common. *P. aucta*, Fab.

b. Thorax narrowed in front, dark green with yellow-red side border.

I. Elytra with rows of fine punctures (the outermost broken into a series of connected punctures), without any yellow longitudinal line on disc.

Oblong, flatly arched. Dark green, side margin of thorax and of elytra broadly yellow-red. L. 1¼—1½ l. Common. *P. marginella*, Lin.

II. Elytra with rows of coarse punctures (the outermost broken into a series of connected punctures toward apex only), with a curved yellow-red longitudinal line on disc.

Ovate, convex, dilated behind. Dark green ; side margin of thorax and of elytra, also a line on disc of latter yellow-red. L. 1½—2 l. Not uncommon. *P. hannoverana,* Fab.

B. Thorax as long as broad.

   a. Elytra black-green, with side margin and a straight longitudinal line on disc yellow.

Black-green ; side margin of thorax and elytra, line on disc of latter, base of femora and the tibiae yellow. Elytra with rows of coarse punctures. L. 2—2¼ l. Common.

P. phellandrii, Lin.

   b. Elytra unicolorous blue.

Blue ; apex of abdomen yellow-red. Elytra with rows of fine punctures. L. 1¾—2 l. Common.

P. beccabungae, Ill.

GALERUCIDAE.

A. Third joint of antennae longer than fourth.

   a. Elytra scarcely longer than broad.

Elytra separately rounded or nearly so at apex ; upper-side of body bare or only slightly hairy.

Adimonia, Laich.

   b. Elytra at least half as long again as broad.

Thorax with a depression on each side ; elytra jointly rounded at apex ; upper-side of body closely punctured, with fine silvery pubescence.   *Galeruca,* Fab.

B. Third joint of antennae shorter than fourth.

   a. Thorax distinctly bordered at base.

Thorax double as broad as long, apex slightly emarginate ; elytra about half as long again as broad, dilated behind.

Agelastica, Redt.

   b. Thorax not, or only indistinctly bordered at base.

    I. Sides of thorax straight.

Thorax half as broad again as long ; elytra double as long as together broad, somewhat blunted at apex.

Phyllobrotica, Redt.

    II. Sides of thorax rounded.

Thorax more than half as broad again as long, apex

straight ; elytra double as long as broad, apex of each more
rounded without than within.   *Luperus*, Geoffr.

Adimonia.

A. Elytra black.

a. Third joint of antennae about half as long again as
second.

Ovate. Black ; upper-side coarsely and deeply punctured ; thorax almost twice as broad as long, tolerably
straight at sides, narrowed before middle, disc shallowly
impressed longitudinally and on each side. L. 4 l. Common.   *A. tanaceti*, Lin.

b. Third joint of antennae only slightly longer than
second.

Oblong-ovate, slightly convex. Pitch-black, shiny ;
under-side with short, scanty, stiff pubescence. Similar to
A. tanaceti but less convex, with antennae shorter, third
joint not much longer than second, punctuation not
wrinkled, thorax broadly but distinctly channeled, less
deeply and not closely punctured, with sides and each side
of apex less deeply impressed. L. 2¾—3 l. Scarce.
                                 A. oelandica, Boh.

B. Elytra unicolorous yellow-brown.

Head, under-side, antennae (except base) and legs (except
tibiae) black ; thorax and elytra pale yellow-brown.
Thorax with two depressions in middle and one on each
side, these depressions usually dark in colour ; elytra
almost jointly rounded but with sutural apical angle obtuse.
L. 2¼—2½ l. Common.   *A. capreae*, Lin.

C. Elytra gray, with suture black-brown.

Oblong, rather convex. Black ; thorax and elytra pale
gray, suture of latter black-brown ; base of antennae,
femora and base of tibiae often reddish. Similar to *A.
capreae*, but with forehead shiny, diffusely but strongly
punctured, facial tubercles polished and well defined behind,
thorax more shiny, with sides more diffusely punctured,
and second and third abdominal segments of male more
closely covered with longer hairs. L. 2—2¾ l. Not uncommon.   *A. suturalis*, Th.

D. Elytra red.

Red ; eyes, apex of antennae, breast and abdomen (except

apex), often scutellum and rarely part of legs black.
Thorax much more than twice as broad as long, sides
rounded, disc coarsely, scantily punctured, with a depression on each side; elytra jointly rounded at apex, sutural
apical angle right angled, their disc sometimes dark. L.
2—2¼ l. Common. *A. sanguinea*, Fab.

Galeruca.

A. Upper-side distinctly punctured.

a. Sutural apical angle of elytra obtuse or rounded.

Upper-side lighter or darker yellow-brown; thorax yellowish; au oblong spot on thorax, vertex of head, scutellum
and shoulders of elytra black; under-side black or pitch-
brown, apex of abdomen and the legs yellow-brown. Forehead with a fine central furrow and without prominences.
L. 2¼—2½ l. Common. *G. lineola*, Fab.

b. Sutural apical angle of elytra forming a sharply
pointed tooth.

I. Thorax dull, as hairy as elytra and with large, distinct punctures throughout, broadest in middle,
evenly narrowed before and behind.

1. Thorax with black central furrow, posterior
angles blunt.

Upper-side lighter or darker yellow-brown; vertex of
head, central furrow of thorax, scutellum and often
shoulders black; under-side black, with prosternum, legs
and anus yellow-brown. Elytra sometimes with a blackish
lateral streak. L. 1¾—2 l. Common.
G. calmariensis, Lin.

2. Thorax entirely yellow, posterior angles distinctly prominent.

Very similar to *G. calmariensis*, but with upper-side
lighter; elytra brownish-yellow; with yellow border.
L. 1¾ l. Common. *G. tenella*, Lin.

II. Thorax shiny, almost bare, punctured in lateral
depressions only, broadest before middle, much
more narrowed behind than before.

1. Thorax with three feeble black spots; length 2
lines.

Black; mouth, thorax (except three feeble black spots), base of joints of antennae and legs (except the slightly darker base of femora) brownish-yellow; elytra dark yellow-brown, lighter at sides. Thorax with a feeble central furrow, usually divided into two depressions. Common.

G. sagittariae, Gyll.

2. Thorax with three distinct black spots; length 2¾ lines.

Similar to *G. sagittariae*, but with spots on thorax distinct, elytra blackish with brownish-yellow sides, rather longer and with sutural apical angle more strongly produced; femora black at base. Rather common.

G. nympheae, Lin.

B. Upper-side very finely shagreened, with scarcely visible punctuation.

Brown, upper-side with yellowish-gray pubescence; a frontal spot, central furrow and sides of thorax and shoulders of elytra blackish. Sides of thorax angularly dilated; sutural apical angle of elytra rounded. L. 2½ l. Rather common.

G. viburni, Payk.

Agelastica.

A. Thorax violet or blue.

Upper-side violet or blue; under-side black-blue; antennae, scutellum, tibiae and tarsi black. Thorax with rounded posterior angles and without depressions on disc; scutellum pointed triangular. L. 2½—2¾ l. Very rare.

A. alni, Lin.

B. Thorax reddish-yellow.

Reddish-yellow; eyes, antennae and scutellum black; vertex of head and the elytra green or blue-green. Thorax with obtuse posterior angles and a depression on each side of disc; scutellum with blunt apex. L. 2½ l. Common.

A. halensis, Lin.

Phyllobrotica.

Yellow; eyes, vertex of head, a point at base of each elytron and a spot before apex, breast and abdomen black. Punctuation fine. L. 2¾—3 l. Rather common.

P. quadrimaculata, Lin.

Luperus.

A. Second and third joints of antennae equal in length; elytra yellow, with black margins.

Black; thorax (except all or part of base) and elytra (except suture, side and apical margins) pale yellow; base of antennae and tibiae yellow-brown. L. $1\frac{1}{4}$—$1\frac{1}{2}$ l. Rather common. *L. circumfusus*, Marsh.

B. Third joint of antennae much longer than second; elytra unicolorous.

a. Thorax black.

Black, shiny; base of antennae and legs (except base of femora) reddish-yellow. Disc of thorax impunctate; elytra very finely, scarcely visibly punctured. L. 2—$2\frac{1}{4}$ l. Common. *L. betulinus*, Fourc.

b. Thorax reddish-yellow.

Black, shiny; thorax, base of antennae and legs (generally except base of femora) reddish-yellow. Thorax impunctate; elytra finely punctured. Male with antennae much longer than body and eyes very large. L. $1\frac{3}{4}$—2 l. Moderately common. *L. flavipes*, Lin.

HALTICIDAE.

A. Body oval or oblong oval.

a. Posterior tarsi inserted at apex of tibiae.

I. First joint of posterior tarsi shorter than half the tibia.

1. Posterior tibiae without tooth on outer-side.

A A. Thorax with a distinct transverse furrow at base.

a a. Elytra confusedly punctured or smooth.

A a. Basal furrow on thorax not bounded at each end by a longitudinal impression.

Posterior femora long and not much thickened; tibiae without visible apical spine. *Haltica*, Ill.

B b. Basal furrow on thorax bounded at each end by a longitudinal impression.

Posterior femora not much thickened; joints of an-

tennae usually shorter than in *Haltica*.
Hermaeophaga, Foudr.

b b. Elytra with punctured striae.

Basal furrow on thorax bounded at each end by a longitudinal impression ; posterior femora moderately thickened ; tibiae without visible apical spine.
Crepidodera, Chevr.

B B. Thorax without basal furrow.

a a. Elytra with punctured striae.

A a. Head sunk in thorax ; latter not narrowed behind.

Forehead with a deep, curved furrow ; thorax with a small oblique impression on each side of base; elytra generally cylindrical, punctured striae strong ; legs not very long ; posterior tibiae channeled.   *Mantura*, Steph.

B b. Head prominent ; thorax evenly narrowed before and behind.

A 1. Base of thorax with a small oblique impression on each side.

Forehead without granulation ; elytra oval; legs long ; posterior tibiae not channeled.   *Podagrica*, Chevr.

B 1. Base of thorax without oblique impressions.

Similar to *Podagrica*.   *Batophila*, Foudr.

b b. Elytra confusedly punctured or smooth.

A a. Forehead with two small tubercles above base of antennae ; thorax tolerably convex ; rather indistinctly punctured.

Margins of clypeus smooth ; vertex of head smooth or indistinctly punctured ; elytra rather scantily punctured.
Aphthona, Chevr.

B b. Forehead without tubercles ; thorax not very convex, tolerably strongly punctured.

Margins of clypeus strongly granulate ; vertex of head more or less strongly punctured ; elytra rather closely punctured.   *Phyllotreta*, Chevr.

2. Posterior tibiae toothed in middle of outer-side.

Head sunk in thorax; forehead with curved transverse furrow; thorax without basal furrow.

Plectroscelis, Chevr.

II. First joint of posterior tarsi as long as, or longer than half the tibia.

Head prominent; thorax somewhat narrow, without impressions, more or less punctured; elytra confusedly and more or less strongly punctured. *Thyamis,* Steph.

b. Posterior tarsi inserted above apex of tibiae.

Antennae with ten joints; forehead with a depression or furrow; thorax narrowed in front, posterior angles not rounded; elytra with regular rows of punctures; posterior tibiae cut off obliquely and produced at apex; first joint of posterior tarsi long. *Psylliodes,* Latr.

B. Body hemispherical.

a. Elytra with punctured striae.

I. Antennae scarcely thickened toward apex.

Forehead with a curved furrow and two small tubercles; vertex of head punctured; thorax transverse, without basal furrow; posterior tibiae channeled.

Apteropeda, Chevr.

II. Antennae with three distinctly larger apical joints.

Forehead with two oblique furrows, crossing each other; thorax transverse, without basal furrow; posterior tibiae not channeled. *Mniophila,* Steph.

b. Elytra confusedly punctured.

Antennae threadlike; clypeus entire; posterior tibiae not channeled. *Sphaeroderma,* Steph.

Haltica.

A. Body flattened.

a. Body oval.

Golden-green, very shiny; head and legs black; underside bronze-black. Thorax short, finely and very closely punctured; elytra very closely, distinctly punctured, with traces of longitudinal inequalities, especially at sides. L. 2¼ l. Common. *H. coryli,* All.

b. Body oblong.

Violet, shiny, not very convex. Frontal tubercles large, oblique sub-angular line scarcely well-defined; facial ridge obtuse; elytra finely punctured, scarcely dilated toward apex, which is only slightly inclined; third joint of antennae slightly longer than second. L. 1¾ l. Rather common.   *H. consobrina*, Kuts.

B. Body convex, oval.

a. Lateral margin of thorax forming tooth at anterior angle.

Green or blue-green; under-side and legs greenish-black. Thorax broader than long, very finely punctured; elytra closely and rather strongly punctured. L. 2½ l. Rather common.   *H. criceti*, All.

H. longicollis, All. is smaller, with punctuation finer; anterior tarsi much dilated.

b. Anterior angles of thorax without tooth.

I. Sides of thorax nearly straight; elytra strongly punctured.

Green, shiny; antennae black; under-side greenish-black. Thorax short, very finely and indistinctly punctured. L. 1½—2 l. Common.   *H. pusilla*, Duft.

The variety *H. montana*, Foudr. is dark blue, rather shorter in proportion.

II. Sides of thorax slightly rounded; elytra not very strongly punctured.

Very similar to *H. pusilla*, but with broader and more distinctly punctured thorax; elytra rather broader and more convex. L. 1¾ l. Common.   *H. helianthemi*, All.

Hermaeophaga.

Short, ovate, convex. Black-blue; under-side blackish; antennae black-brown. Punctuation very fine. L. 1—1¼ l. Common.   *H. mercurialis*, Fab.

Crepidodera.

A. Under-side reddish-yellow.

a. Elytra with rows of double punctures.

Oblong-ovate. Reddish-yellow-brown. Forehead with

two granules; thorax punctured, impression punctured at base only. L. 2—2½ l. Common. *C. transversa*, Marsh.

b. Elytra with rows of simple punctures.

Oblong-ovate. Red-yellow. Forehead with two granules; thorax very finely punctured; elytra rather deeply punctured. L. 1½ l. Common. *C. ferruginea*, Scop.

B. Under-side wholly or chiefly black, bronze-black or greenish-black.

    a. Elytra wholly or partly black, blue, green or bronze.

        I. Elytra not pubescent.

            1. Elytra unicolorous.

            A A. Thorax reddish.

Yellow-red; elytra blue or green; breast and abdomen black. Thorax broadest in middle, impunctate; elytra with regular rows of not very deep punctures, reaching apex. L. 1½ l. Rather common. *C. rufipes*, Lin.

            B B. Thorax concolorous with elytra or golden-green.

                a a. Length 1¾—2 lines.

                    A a. Elytra with rows of fine punctures, the inner rows irregular; antennæ red-yellow at base, dark-brown at apex.

Oblong-ovate. Head and thorax golden-green; elytra blue; under-side black, legs red-yellow, posterior femora bronze-black. Thorax finely and diffusely punctured. L. 1¾ l. Not uncommon. *C. nitidula*, Lin.

                    B b. Elytra with regular rows of strong punctures; antennae almost entirely reddish.

Similar to *C. aurata*, but more robust; thorax more convex, with shallower basal furrow, sides more rounded, posterior angles obtuse; elytra broader and more convex. L. 1¾—2 l. Common. *C. helxines*, Lin.

                b b. Length 1—1½ line.

                    A a. Antennae with first five joints reddish, rest brown.

Ovate. Green or violet; thorax coppery; legs reddish, posterior femora brown. Thorax deeply punctured, with

interstices again very finely punctured ; elytra with regular
rows of deep punctures, interstices convex, very finely
wrinkled.   Common.   *C. aurata*, Marsh.

B b. Antennae with first four joints reddish,
rest black.

Oblong.   Coppery-green ; legs reddish, posterior femora
bronze.   Thorax deeply punctured, with interstices more
or less wrinkled ; elytra with regular rows of punctures,
interstices scarcely visibly punctured.   Not uncommon.
C. chloris, Foudr.

2. Elytra bronze, with apex yellow.

Short ovate, convex, very shiny.   Head and thorax
bronze ; breast and abdomen bronze-black, apex of latter
reddish ; antennae red-yellow at base, black at apex ; legs
pale red-yellow, posterior femora pitchy at apex.   Thorax
very finely punctured, basal furrow shallow ; elytra with
regular rows of rather deep punctures.   L. 1 l.   Common.
C. modeeri, Lin.

II. Elytra pubescent.

1. Elytra unicolorous black.

Ovate, convex.   Black, shiny ; base of antennae and legs
reddish-yellow.   Thorax strongly and closely punctured ;
elytra with regular rows of rather deep punctures.   L. ¾—1 l.
Not uncommon.   *C. pubescens*, E.H.

2. Elytra black, with two red spots on each.

Very similar to *C. pubescens*, but with two red spots on
each elytron (basal one sometimes absent) and with scantier
pubescence.   L. ¾ l.   Rather common.   *C. atropae*, Foudr.

b. Elytra pale reddish.

I. Antennae pale red-yellow at base, grayish-brown
at apex.

Ovate, convex.   Pale red-yellow ; elytra yet lighter ;
breast and abdomen pitch-black ; eyes black.   Thorax
very convex, very finely punctured ; elytra much broader
at base than thorax, with rows of rather deep punctures,
effaced before apex.   L. ¾—1 l.   Common.
C. salicariae, Payk.

II. Antennae red-yellow, with apical joints darker
at apex and last joint black.

Ovate, convex. Pale red-yellow; breast, abdomen, and eyes black. Thorax not very convex, finely punctured; elytra a little broader at base than thorax, with regular rows of tolerably strong punctures, effaced before apex. L. 1 l. Moderately common. *C. ventralis*, Ill.

Mantura.

A. Length 1¼—1½ lines.

   a. Elytra dark blue or greenish-blue, with apex reddish; front pairs of femora red-yellow.

Elongate ovate, very convex. Head and thorax greenish-bronze; under-side black; antennae reddish at base, black at apex; legs red-yellow, with posterior femora greenish-black. Forehead with large punctures; thorax strongly punctured. Common. *M. rustica*, Lin.

   b. Elytra entirely blackish-blue; all femora black.

Oblong-ovate, convex. Upper-side blackish-blue; under-side black; antennae red-yellow at base, black at apex; legs (except femora) reddish. Head and thorax finely and closely punctured, latter and elytra a little broader than in *M. rustica*. L. 1¼ l. Not very common.

                          M. obtusata, Gyll.

B. Length 1 line.

   a. Elytra bronze or coppery (apex sometimes yellow-red); front pairs of femora reddish.

Upper-side bronze or coppery; under-side black, with bronze reflection; antennae red-yellow at base, black at apex; legs reddish, with posterior femora pitchy. Much shorter and more oval than *M. rustica*, thorax more finely and closely punctured, base of elytra scarcely broader than that of thorax. Not common. *M. chrysanthemi*, E. H.

   b. Elytra bronze-green or blue; all femora pitch-black.

Head and thorax green, or bronze-green; under-side black; antennae red-yellow at base, black at apex; legs (except femora) red. More cylindrical than *M. chrysanthemi*, thorax a little narrower in front, and more closely punctured, first two rows of punctures on elytra slightly confused at base. Rather common. *M. Matthewsi*, Curt.

Podagrica.

A. Punctures on elytra distinctly arranged in rows.

Oblong. Head and thorax reddish ; elytra bronze-green ; breast reddish, abdomen and legs black ; first four joints of antennae red, rest black-brown. Elytra with rows of tolerably strong punctures and very finely punctured interstices. L. 1¼—1½ l. Common. *P. fuscipes,* Fab.

B. Punctures on elytra not distinctly arranged in rows.

Ovate. Head, thorax and legs red ; elytra blue or greenish-blue ; under-side (except of head and thorax) black ; antennae red at base, black at apex. Elytra very closely and finely punctured. L. 1½—2 l. Common. *P. fuscicornis,* Lin.

Batophila.

A. Upper-side black.

Elytra sometimes with bluish or bronze reflection ; under-side black ; antennae and legs yellow-red. Elytra with regular rows of deep punctures, interstices impunctate. L. ¾ l. Common. *B. rubi,* Payk.

B. Upper-side green, slightly bronzed.

Under-side black ; antennae and legs yellow-red, last two or three joints of former sometimes rather darker. Elytra with regular rows of deep punctures, interstices impunctate. L. ¾ l. Common. *B. aerata,* Marsh.

Aphthona.

A. Elytra red-yellow, with suture black.

a. Head red-yellow.

Ovate, not very convex. Thorax red-yellow ; under-side black, with sides of prosternum pale ; antennae black, with first four or five joints red-yellow ; legs red-yellow, tarsi brown, posterior femora black at apex. L. 1—1¼ l. Common. *A. lutescens,* Gyll.

b. Head black.

Ovate, not very convex. Thorax yellow-red ; under-side black ; antennae red-yellow, brown at apex ; legs red-yellow. L. ¾ l. Rather common. *A. nigriceps,* Redt.

B. Elytra black, green or blue.

a. Shoulders prominent.

I. Length 1—1½ lines.

1. Front pairs of femora entirely red-yellow ; posterior femora blackish at apex only.

Oblong. Upper-side blue, shiny ; under-side black ; first five joints of antennae red-yellow, rest brown-black ; legs (except apex of posterior femora) red-yellow. Thorax scarcely visibly punctured ; elytra finely, confusedly punctured, interstices somewhat wrinkled, sutural apical angle almost a right angle, under-side indistinctly punctured. L. 1—1½ l. Common. *A. coerulea,* Payk.

2. Front pairs of femora red-yellow, with a pitch-brown spot in middle ; posterior femora entirely pitch-brown.

Ovate. Head black ; thorax and elytra violet-black ; under-side black ; antennae red-yellow, apex brown ; tibiae and tarsi red-yellow. Thorax impunctate ; elytra very finely, confusedly punctured, sutural apical angle almost a right angle. L. 1—1¼ l. Rather common.

A. venustula, Kuts.

II. Length ½—¾ line.

1. Sides of thorax scarcely visibly punctured.

Short reversed ovate. Upper-side dark blue ; under-side black ; first five joints of antennae red-yellow, rest dark brown ; legs yellow-red, posterior femora often brown. Thorax scarcely visibly punctured ; elytra rather strongly punctured, in irregular rows at base, confusedly behind middle ; under-side strongly punctured. L. ½—¾ l. Common. *A. cyanella,* Redt.

2. Sides of thorax distinctly punctured.

Ovate. Upper-side bronze-green, elytra varying to bluish-black ; under-side bronze-black ; antennae red-yellow at base, last five joints brown ; legs red-yellow, posterior femora often red-brown. Punctuation of thorax scarcely visible on disc ; elytra distinctly punctured, with an admixture of wrinkles, sutural apical angle obtuse. L. ¾ l. Moderately common. *A. hilaris,* Steph.

b. Shoulders rounded.

I. Thorax nearly twice as broad as long ; punctuation very fine.

Oblong-ovate. Black, shiny; antennae with first six joints and base of seventh yellow-red, rest black; legs yellow-red, posterior femora black. Elytra strongly punctured, separately rounded at apex; joints five to ten of antennae dilated at apex, about equal in length. L. ¾ l. Not very common.   *A. atratula*, All.

II. Thorax nearly square; punctuation distinct.

Oblong-ovate. Green, slightly bronzed, shiny; underside black; antennae red-yellow, last three or four joints somewhat brown. Elytra strongly punctured, separately rounded at apex; joints five to ten of antennae not dilated at apex, fifth rather longer than rest. L. ¾—1 l. Common.
A. herbigrada, Curt.

Phyllotreta.

A. Elytra unicolorous.

a. Upper-side bronze-brown or bronze-green.

I. Antennae with first four joints red-yellow, rest black.

Oblong. Greenish-black or bronze-brown; under-side and legs black. Thorax somewhat broader than long; elytra almost parallel-sided, separately rounded at apex; punctuation somewhat strong, confused. Male with fourth joint of antennae strongly and fifth somewhat dilated. L. 1 l. Common.   *P. nodicornis*, Marsh.

II. Antennae unicolorous black.

Oblong-ovate. Bronze-green (sometimes blue); underside and legs black. Thorax transverse; elytra slightly rounded at sides, separately rounded at apex; punctuation fine, confused. L. ¾—1 l. Moderately common.
P. lepidii, E. H.

b. Upper-side black.

1. Antennae unicolorous black.

Oblong. Black, sometimes bluish-black; tarsi brown. Thorax transverse; elytra slightly rounded at sides, apex obtusely rounded; punctuation of thorax rather strong, of elytra feebler and confused. Male with third, fourth and fifth joints of antennae dilated. L. ¾—1 l. Rather common.   *P. melaena*, Ill.

11. Antennae with base red-yellow.

1. Elytra punctured in rows.

A A. Upper-side deep black.

Oblong-ovate. Under-side and legs black; joints and tarsi somewhat reddish. Thorax twice as broad as long, sides rounded; apex of elytra separately rounded; punctuation strong. L. ¾—1 l. Common. *P. atra*, Payk.

B B. Upper-side black, with greenish or bluish reflection.

Similar to *P. atra*, but with base of antennae paler, posterior angles of thorax rather more nearly right angles and rows of punctures on elytra more regular. L. ¾—1 l. Not uncommon. *P. obscurella*, Ill.

2. Elytra confusedly punctured.

Elongate, elliptic. Black or bronze-black; under-side and legs black, base of posterior tibiae and the tarsi slightly reddish. Thorax transverse, sides slightly rounded; elytra very obtusely rounded at apex; punctuation rather fine. L. ¾—1 l. Not common. *P. punctulata*, Marsh.

B. Elytra black and yellow.

a. Each elytron with a longitudinal yellow band.

I. Yellow band on elytra not strongly emarginate in middle of outer-side, and its inner margin nearly straight.

1. Yellow band on elytra abruptly sloped inward at base.

Oblong-ovate. Black; head and thorax with a greenish reflection; elytra as above; base of antennae, apex of anterior femora, knees of hinder pairs, under-side of tibiae and the tarsi reddish. Elytra punctured strongly and in rows at base, more feebly and confusedly at apex. L. ¾—1 l. Rather common. *P. vittula*, Redt.

2. Yellow band on elytra gradually sloped inward at base.

A A. Thorax black; tibiae red-yellow at base only.

a a. Body flat.

Oblong-ovate. Black; bands on elytra pale yellow; first three joints of antennae, base of tibiae and the tarsi red-yellow. Thorax and elytra rather more strongly punctured than in *P. nemorum*, punctures on latter in rows on middle only and the yellow band narrower, slightly curved toward suture at base and apex and ending farther from apex. L. 1 l. Common. *P. undulata*, Kuts.

b b. Body rather convex.

Oblong-ovate, somewhat convex, shiny. Black; punctulate; elytra nearly ovate, with a narrow, longitudinal yellow streak on each, the outer-side of which is emarginate in middle, the inner-side nearly straight, and which is recurved toward suture at apex only; base of antennae and knees dusky yellow. L. 1—1¼ l. Rare.

P. flexuosa, Kuts.

B B. Thorax black, generally with greenish reflection; tibiae entirely red-yellow.

Oblong-ovate. Head, under-side, antennae (except first three joints) and femora black. Punctuation close and deep, in rows on elytra, the yellow band on which turns inward at apex, but does not reach apex of elytra. L. 1 —1¼ l. Common. *P. nemorum*, Lin.

II. Inner margin of yellow band on elytra concave, outer margin strongly emarginate in middle.

1. Only base of tibiae red-yellow.

Oblong-ovate. Black; elytra as above; first three joints of antennae and base of tibiae yellow-red. Punctuation strong, on elytra generally in rows, especially at base. Male with fourth and fifth joints of antennae dilated. L. 1 l. Common. *P. sinuata*, Steph.

2. Front pairs of legs entirely red-yellow.

Ovate. Black; elytra as above; first three joints of antennae and the legs (except posterior femora) red-yellow. Thorax more finely punctured, body rather more oval and more convex than in *P. sinuata*, yellow band on elytra broad. Male with fifth joint of antennae twice as broad as second. L. 1—1¼ l. Rather common.

P. ochripes, Steph.

b. Each elytron with two yellow spots (sometimes connected by a line).

I. Length 1—1¼ line.

Oval, somewhat convex. Black; elytra as above; first two or three joints of antennae and sometimes the knees and base of tibiae reddish; tarsi brown. Punctuation of thorax feeble, that of elytra stronger and sometimes in rows. Rather common. *P. tetrastigma*, Com.

II. Length ¾ line.

Oval, shorter and more convex than *P. tetrastigma*. Black; elytra as above; first three joints of antennae red-yellow; tibiae and tarsi brownish-red. Punctuation close, that of elytra stronger than that of thorax. Common.

P. brassicae, Fab.

Plectroscelis.

A. Elytra with regular rows of punctures throughout.

Ovate, not very convex. Upper-side bronze-green; under-side black; antennae reddish at base, brown at apex; femora black, tibiae and tarsi dark reddish. Elytra strongly punctured, interstices smooth; posterior tibiae toothed. L. ¾ l. Common. *P. concinna*, Marsh.

B. Disc of elytra with irregular rows of punctures.

a. Body oblong.

Oblong-ovate, convex, not very shiny. Black-blue; base of antennae rust-red, first joint spotted with brown at base; tibiae and tarsi reddish, more or less brown. Vertex of head and thorax finely punctured, latter transverse; elytra oblong-ovate, finely punctured in rows, confused toward suture, humeral callosity distinct; winged. Punctuation finer than in *P. Sahlbergi*, that on elytra more confused toward suture, body less rounded at sides, thorax rather longer and more parallel-sided. L. 1 l. Moderately common. *P. subcoerulea*, Kuts.

b. Body ovate.

I. Upper-side dark blue or green; thorax more convex.

Ovate, convex, rather shiny. Dark blue or green; base of antennae reddish, first joint spotted with brown at base; tibiae and tarsi reddish. Vertex of head and thorax closely and deeply punctured; elytra ovate, with rows of deep punctures, dorsal rows more or less confused, humeral callosity distinct; winged. L. 1 l. Scarce.

P. Sahlbergi, Gyll.

II. Upper-side brownish or greenish-bronze ; thorax less convex.

Brownish or greenish-bronze, above ; under-side bronze-black ; first four joints of antennae entirely red-yellow, rest black; femora bronze, tibiae and tarsi red-yellow. Punctuation of thorax and elytra not so strong as in *P. Sahlbcrgi*, the former much more transverse and the rows of punctures on latter not so regular. L. ¾—1 l. Rather common.

P. aridella, Payk.

C. Disc of elytra confusedly punctured.

a. First two joints of antennae spotted with black.

Ovate. Dark bronze above ; under-side and femora bronze-black ; tibiae and tarsi reddish, former generally brownish in middle ; first three joints of antennae reddish, first two spotted with black, rest black. Head and thorax very finely punctured, latter somewhat broader than long ; elytra rather finely punctured. L. 1 l. Rare.

P. aridula, Gyll.

b. First two joints of antennae entirely red-yellow.

Ovate. Bronze-black above ; under-side black ; femora bronze-black, tibiae and tarsi red-yellow ; first four or five joints of antennae entirely red-yellow, rest black. Thorax and elytra punctured rather more strongly than in *P. aridula*. L. 1—1¼ l. Scarce.   *P. confusa*, Boh.

Thyamis.

A. Elytra black, blue or green, with or without lighter markings.

a. Elytra unicolorous.

I. Elytra not very finely punctured.

1. Body rather flat; elytra completely covering abdomen.

A A. Elytra pointed at apex, very closely and rather deeply punctured; length 1¼ line.

Black ; base of antennae and legs (except posterior femora) yellow. Thorax extremely finely punctured, almost smooth; winged. Rare.   *T. nigra*, E. H.

B B. Elytra obtusely rounded at apex, finely punctured, almost in rows; length ¾ line.

Oblong-ovate. Pitch-black ; thorax and elytra dark

bronze; base of antennae and legs (except posterior femora) yellow. Thorax closely and finely punctured. Rather common. *T. obliterata*, Rosenh.

2. Body very convex; elytra not entirely covering abdomen.

Ovate. Black; base of antennae, tibiae and tarsi yellow. Thorax closely and very finely punctured; elytra more closely and deeply punctured than thorax, separately rounded at apex; apterous. L. ¾ l. Not uncommon.

T. anchusae, Payk.

II. Elytra very finely punctured.

Ovate. Upper-side bronze-brown or pitch-brown; underside black; first five joints of antennae red-yellow, rest brown-black; legs dull red-yellow, posterior femora pitch-brown. Punctuation extremely fine, indistinct; winged. L. ¾ l. Moderately common. *T. parvula*, Payk.

b. Elytra with apex lighter.

Ovate, convex. Black, shiny; elytra with a round red spot before apex; first three joints of antennae and anterior tibiae and tarsi red-yellow, hinder pairs of latter darker. Punctuation close and fine. L. 1 l. Rather common.

T. holsatica, Lin.

c. Each elytron with two lighter spots.

I. Posterior tibiae brown.

Oblong oval, convex. Black, shiny, with bluish reflection; elytra with a spot before apex and at shoulder obscure red-yellow; base of antennae and legs red-yellow, posterior femora black, anterior femora until beyond middle, posterior tibiae and apex of all tarsi brown. Elytra closely, not finely, confusedly punctured, sutural angle obtuse; wings very short. Fifth ventral abdominal segment of male with a small tubercle in middle of apex. Elytra more strongly punctured than in *T. anchusae*, with apex less rounded. L. ⅔ l. Not uncommon.

T. absinthii, Kuts.

II. Posterior tibiae yellow-red.

Oblong-ovate, convex. Black, shiny; each elytron with a reddish spot (variable in size) at shoulder and another at apex; base of antennae and legs (except posterior femora) yellow-red. Punctuation rather close, distinct. L. 1¼—1½ l. Moderately common. *T. quadripustulata*, Fab.

d. Elytra with side margins lighter.

Ovate, not very convex.   Black ; thorax yellow-red,
anterior margin often black ; elytra black, with a broad red-
yellow border ; base of antennae and legs (except posterior
femora) reddish.   Thorax smooth, elytra finely punctured.
L. ¾—1 l.   Moderately common.   *T. dorsalis,* Fab.

B. Elytra unicolorous brown, or red-brown.

a. Antennae entirely reddish ; outer margin of elytra
with long curved hairs.

Ovate.   Dark brown or black-brown ; legs reddish, posterior femora rather darker at apex.   Thorax half as broad
again as long ; elytra confusedly punctured, tolerably
strongly at base, gradually more indistinctly toward apex,
where the punctuation is almost effaced.   L. 1—1½ l.
Moderately common.   *T. castanea,* Foudr.

b. Antennae black at apex ; outer margin of elytra
without hairs.

I. Thorax nearly as long as broad ; punctuation of
elytra fine.

Ovate, convex.   Red-brown, rather dull ; mouth, eyes
and apex of antennae black.   Punctuation of thorax fine,
not close, that of elytra more distinct and scattered.   L. 1
l.   Rather common.   *T. brunnea,* Duft.

II. Thorax transverse ; punctuation of elytra strong.

1. Body convex ; length ¾—1 line.

Ovate, convex.   Blackish-brown or red-brown ; base of
antennae and legs red-yellow, posterior femora brown above.
Punctuation of thorax very close, but indistinct.   Rather
common.   *T. lurida,* Scop.

2. Body rather convex ; length ⅔—¾ line.

Oblong-ovate, rather convex, shiny.   Brown or pitch-
brown ; shoulders, outer margin and apex of elytra, base of
antennae and legs paler.   Thorax transverse, very finely
granulate and minutely punctured ; elytra with shoulders
scarcely prominent, distinctly, closely and confusedly
punctured in wrinkles, apices separately obtusely rounded ;
apterous.   L. ⅔—¾ l.   Scarce.   *T. fuscula,* Kuts.

C. Elytra yellowish, with suture darker.

a. Suture generally blackish.

I. Shoulders of elytra obliquely rounded.

1. Elytra jointly or separately rounded at apex.

A A. Thorax much narrower than elytra, scarcely punctured.

Ovate, moderately convex. Head and under-side pitch-black; thorax brown-red; elytra yellow, suture broadly black; first three joints of antennae reddish, rest black; legs reddish, posterior femora pitch-black. Elytra very finely, indistinctly punctured, separately rounded at apex. L. 1 1. Common. *T. fuscicollis*, Steph.

B B. Thorax a little narrower than elytra, distinctly punctured.

a a. Body oblong-ovate.

A a. Antennae not stout; thorax with bronze reflection.

Oblong-ovate, convex. Head and under-side pitch-black; thorax yellow-red, with bronze reflection; elytra grayish, suture black-brown; first four or five joints of antennae red-yellow, rest black; legs yellow, posterior femora pitch-black, generally reddish at base and beneath. Elytra closely and distinctly punctured, separately rounded at apex. L. 1—1¼ 1. Common. *T. atricilla*, Gyll.

B b. Antennae stout; thorax without bronze reflection.

Oblong, apterous. Rust-red; under-side, suture of elytra and sometimes thorax darker; apex of posterior femora broadly, apical joint of tarsi and apex of antennae pitch-black. Similar to *T. atricilla* but rather larger, less regularly oval; head lighter and without bronze reflection; elytra with more prominent shoulders and punctured rather more strongly, in rows toward base, sutural angle less obtuse; apical spine of posterior tibiae much longer; posterior tarsi longer; antennae very stout, first joint as long as next two together, second and third short, latter the longer, fourth and fifth nearly equal, each much longer than third, succeeding joints gradually rather shorter and broader. L. 1—1¼ 1. Rare. *T. distinguenda*, Rye.

b b. Body ovate.

A a. Elytra punctured in rows at base.

Ovate. Brownish-yellow ; elytra lighter, generally with hinder part of suture, sometimes with disc dark ; posterior femora often pitch-black. Thorax distinctly and diffusely punctured ; elytra punctured distinctly and almost in rows at base, less deeply toward apex. Punctuation of thorax and elytra stronger than in *T. melanocephala*, shoulders of latter broader, antennae stouter. L. 1½ l. Rare.

T. patruelis, All.

B b. Elytra not punctured in rows.

Ovate, convex, shiny. Yellow-red ; eyes black ; head and suture of elytra sometimes blackish ; under-side and apex of antennae pitchy ; more or less of base of posterior femora and sometimes the apex of tibiae and tarsi pitchy ; winged. Darker than *T. verbasci*, thorax more distinctly, elytra more strongly and less closely punctured, second and third joints of antennae equal in length, and spur at apex of posterior tibiae very much shorter and scarcely percep- tibly curved. L. 1⅛—1⅓ l. Rare.   *T. agilis*, Rye.

2. Elytra almost pointed at apex.

A A. Elytra coarsely and strongly punctured.

Oblong-ovate, rather convex. Head, scutellum, breast and abdomen pitch-black ; thorax yellow-red ; elytra pale red-yellow, suture narrowly brown ; base of antennae, an- terior legs and posterior tarsi pale red-yellow, posterior femora black-brown, their tibiae brownish. Thorax short, finely punctured ; apical spine of posterior tibiae rather short. L. ¾—1 l. Scarce.   *T. atriceps*, Kuts.

B B. Elytra finely punctured.

Ovate, convex. Head and under-side pitch-black, thorax yellow-red ; elytra pale red-yellow, suture dark brown ; first five or six joints of antennae red-yellow, rest dark brown ; femora black, front pairs generally yellowish at apex ; tibiae and tarsi red-yellow, posterior tibiae darker. Thorax scarcely punctured ; elytra with fine, very close punctuation. L. 1 l. Common.

T. melanocephala, Gyll.

II. Shoulders of elytra not obliquely rounded.

1. Thorax black or pitch-black.

A A. Elytra with suture and outer margin black ; posterior femora red at base.

Ovate, not very convex. Head, thorax and under-side black ; elytra pale red-yellow, suture narrowly and outer margin more broadly brown-black ; first four or five joints of antennae red-yellow, rest black ; legs (except apex of posterior femora) red-yellow. Thorax much narrower than elytra, closely, finely and indistinctly punctured ; elytra rather strongly and distinctly punctured, almost in rows at base. L. 1—1¼ l. Not uncommon. *T. suturalis*, Marsh.

B B. Elytra with all margins black ; posterior femora entirely pitch-black.

Ovate, not very convex. Head, thorax and under-side pitch-black ; elytra pale yellow, bordered with pitch-black ; base of antennae and legs (except posterior femora) red-yellow, apex of former black. Thorax finely but distinctly punctured ; elytra confusedly and distinctly punctured. L. 1 l. Common. *T. nasturtii*, Fab.

2. Thorax yellowish or yellow-red.

A A. Prosternum yellow, rest of under-side black ; body oblong oval.

Not very convex. Head and under-side pitch-black ; thorax and elytra pale yellow, suture black, very narrowly at base and apex ; base of antennae and legs (except posterior femora) red-yellow, apex of former and posterior femora black. Thorax almost smooth ; elytra finely and closely punctured. L. 1¼ l. Moderately common.
T. Foudrasi, Crotch.

B B. Under-side entirely pitch-brown ; body ovate.

Convex. Head and under-side pitch-brown ; thorax and elytra pale red-yellow, suture and sometimes side margin of latter very narrowly pitch-brown ; base of antennae and legs (except posterior femora) red-yellow, apex of former brownish, posterior femora reddish, darker above. Punctuation of thorax fine, that of elytra rather stronger and almost in rows at base. L. ¾ l. Not uncommon.
T. lycopi, Foudr.

b. Suture reddish.

Ovate, convex. Head reddish ; thorax yellow-red ; scutellum dark, often black ; elytra pale red-yellow, suture narrowly reddish ; prosternum reddish, rest of under-side pitch-black ; first four joints of antennae and legs (except posterior femora) red-yellow, apex of former black ; posterior femora reddish, with a blackish spot above. Punctuation fine ; posterior femora and tibiae long. L. 1¼—1½ 1. Rather common. *T. femoralis*, Marsh.

D. Elytra unicolorous pale yellow, yellow or red.

    a. Elytra not gaping at apex.

      I. Antennae darker at apex than at base.

        1. Punctuation strong.

          A A. Head pitch-black.

Oblong-ovate, convex. Head and under-side pitch-black ; thorax and elytra pale red-yellow ; antennae red-yellow at base, black at apex ; legs red-yellow, posterior femora pitch-brown. Elytra nearly pointed at apex. L. 1 1. Rather common. *T. ballotae*, Marsh.

          B B. Head reddish.

            a a. Punctuation of elytra distinctly in rows.

Oblong-ovate. Reddish ; labrum brown. Thorax finely punctured ; elytra more closely and strongly punctured, most of the punctures arranged in rows and mixed with wrinkles. Apterous. L. ¾ 1. Rare. *T. cerina*, Foudr.

            b b. Punctuation of elytra confused, or only in indistinct rows.

Ovate, convex, shiny. Head rust-red, vertex darker, labrum pitch-black ; thorax reddish ; elytra, base of antennae and the legs red-yellow ; abdomen black, with last segment and pygidium rust-red. Thorax transverse, punctured in wrinkles ; elytra ovate, with shoulders prominent, punctuation strong, confused or in indistinct rows, apices scarcely jointly rounded ; apical spine of posterior tibiae long ; apterous. L. 1 1. Rare. *T. Waterhousei*, Kuts.

        2. Punctuation fine.

          A A. Abdomen pitch-black.

            a a. Posterior femora black at apex, reddish at base.

Ovate, convex. Head and under-side pitch-black ; thorax pitch-brown, disc reddish ; elytra pale red-yellow ; antennae red-yellow at base, brown at apex ; legs yellow-red, posterior femora more or less brown-black on upper-side and apex. L. ¾ l. Common. *T. pusilla,* Gyll.

> **b b.** Posterior femora entirely red-yellow.

Head dark brown ; thorax and elytra pale red-yellow, anterior margin of former pitch-brown ; prosternum red-yellow, rest of under-side pitch-black ; first four or five joints of antennae pale red-yellow, rest dark brown; legs pale red-yellow, posterior femora slightly darker. Punctuation indistinct. L. ¾—1 l. Rare. *T. Reichei,* All.

> **c c.** Posterior femora entirely pitch-brown.

Oblong-ovate, convex. Head and under-side pitch-black, front of former reddish ; thorax yellowish-red ; elytra red-yellow ; antennae red-yellow at base, brownish at apex ; legs (except posterior femora and sometimes also apex of posterior tibiae) red-yellow. Thorax rather indistinctly punctured ; elytra finely but visibly punctured. L. 1 l. Scarce. *T. medicaginis,* All.

> ***B B.*** Abdomen brown or yellow.

> > ***a a.*** Length 1½—2 lines.

> > > ***A a.*** Elytra gray-yellow or yellow.

> > > > ***A 1.*** Thorax scarcely narrower than elytra ; posterior femora red-brown.

Oblong-ovate, convex, shiny ; upper-side gray-yellow, head darker ; prosternum gray-yellow, rest of under-side red-brown ; antennae grayish, apex darker ; legs pale, posterior femora red-brown, darker above and at apex. Punctuation of thorax scarcely visible ; that of elytra very fine. L. 1½—2 l. Common. *T. verbasci,* Panz.

The variety *T. thapsi* has the head and under-side pitch-brown, thorax and elytra pale red-yellow, latter with suture narrowly black and antennae (except base) black.

> > > > ***B 1.*** Thorax much narrower than elytra ; posterior femora pale red-yellow.

Ovate, not very shiny, rather less convex than *T. verbasci.* Pale yellow ; mouth, eyes and apex of antennae blackish. Punctuation of thorax very indistinct, that of elytra very fine. L. 1½ l. Common. *T. tabida,* Panz.

> > > ***B b.*** Elytra red.

Ovate, shiny. Yellowish or reddish ; elytra red ; apex

of antennae and generally under-side brownish.   First joint
of posterior tarsi rather longer than in *T. tabida,* punctuation of elytra stronger.   L. 1½ l.   Not uncommon.

T. rutila, Ill.

b b. Length ¾—1¼ lines.

A a. Elytra punctured at apex; insects
winged.

A 1. Apex of posterior femora black.

Ovate, not very convex.   Pale yellow ; mouth, eyes and
apex of posterior femora black ; breast and apex of antennae
brown.   Thorax smooth ; elytra very finely punctured ;
forehead distinctly ridged.   L. 1—1¼ l.   Common.

T. ochroleuca, Marsh.

B 1. Posterior femora entirely yellowish.

Oblong-ovate, rather flat, shiny.   Whitish-yellow ; head,
breast and abdomen pale rust-red ; mouth pitchy; apex of
antennae and last joints of tarsi (with claws) brownish.
Thorax transverse, rather broad, not or scarcely punctured ;
elytra ovate, transparent, shoulders moderately prominent,
very finely and indistinctly confusedly punctured, apices
nearly separately rounded ; apical spine of posterior tibiae
short ; winged.   L. ¾—⅞ l.   Not common.

T. gracilis, Kuts.

The variety *T. Poweri,* All. has the vertex of head,
scutellum, sutural margins of elytra, breast and abdomen
brown ; joints of antennae rather longer, apical joints
darker, punctuation of thorax closer and (as also that of
elytra) more distinct ; male with depression on last segment of abdomen rounded.

B b. Elytra impunctate at apex ; insects
apterous.

A 1. Shoulders of elytra rounded, apex
of posterior femora pitch-brown.

Oblong-ovate, convex.   Pale red-yellow, under-side
slightly darker ; mouth and apex of posterior femora pitch-
brown ; apex of antennae dark.   Head and thorax impunctate ; punctuation of elytra scarcely visible, even when
strongly magnified.   L. 1 l. Common.   *T. laevis,* Duft.

B 1. Shoulders of elytra distinctly prominent ; posterior femora entirely reddish.

Ovate, convex. Red-yellow, breast and abdomen rather darker; mouth and last three joints of antennae pitch-black; posterior femora reddish. Thorax and elytra distinctly punctured. L. 1 l. Not common.

T. pellucida, Foudr.

II. Antennae entirely red-yellow.

Ovate, convex. Head and under-side yellow-red; thorax and elytra red-yellow; legs pale red-yellow, posterior femora rather less pale. Punctuation strong; elytra separately rounded at apex, shoulders rounded. L. 1¼ l. Not common.

T. flavicornis, Steph.

b. Elytra gaping at apex.

Ovate, convex. Upper-side red-yellow, elytra slightly paler; under-side yellow-red; antennae yellowish at base, reddish at apex; legs red-yellow, posterior femora reddish. Punctuation strong. L. ¾—1 l. Common.

T. teucrii, All.

Psylliodes.

A. Forehead with a semicircular furrow between eyes or without any furrow.

a. Head dark blue.

I. Punctuation of thorax rather strong at sides, finer on disc.

1. Legs black, posterior femora violet.

Ovate, convex. Dark blue, under-side blackish; base of antennae reddish-yellow. Elytra with regular rows of punctures, interstices indistinctly punctured. L. 1½ l. Not uncommon.

P. dulcamarae, E. H.

2. Legs red-yellow, apex of front pairs of femora brown, posterior femora bronze-black.

Short, ovate. Upper-side blue or greenish-blue; under-side bronze-black; first four joints of antennae red-yellow, rest brown. Elytra with rows of strong punctures, interstices punctured. L. 1¼ l. Moderately common.

P. chalcomera, Ill.

II. Punctuation of thorax very fine and obsolete.

Oblong-ovate, convex. Upper-side dark blue, rarely greenish; under-side black, with some bluish or bronze reflection ; antennae red-yellow at base, brown at apex ; legs pale red-yellow, posterior femora black. Elytra with regular rows of punctures, interstices scarcely punctured. L. 1—1½ l.　Common.　　　　　　　　　　*P. napi*, E.H.

b. Head bronze-green or yellow-red.

　I. Thorax not strongly punctured.

　　1. Under-side wholly or chiefly blackish.

　　A A. Elytra bronze-green or blackish-blue.

　　　a a. Head entirely bronze-green.

Ovate, convex.　Upper-side bronze-green ; under-side bronze-black ; antennae red-yellow at base, pitch-black at apex ; legs red-yellow, posterior femora bronze-black. Punctuation of thorax finer but closer than in *P. chalcomera*, and rows of punctures on elytra deeper, interstices finely punctured.　L. 1½ l.　Rare.　　　　*P. hyoscyami*, Lin.

　　　b b. Head yellow-red, with vertex pitchy-red or dark green.

Ovate, convex, narrowed behind.　Thorax and elytra dark blue or green ; under-side pitch-black ; antennae red-yellow at base, brown at apex ; legs yellow-red, posterior femora black, front pairs of legs sometimes darker.　L. 1¾—2 l.　Common.　　　　　　*P. chrysocephala*, Fab.

The variety *P. nigricollis* has the vertex of head and thorax bronze-green, elytra red-yellow.

　　　c c. Head entirely yellow-red.

Oblong-ovate, convex, narrowed behind.　Head, thorax, prosternum, base of antennae and legs yellow-red ; elytra dark blue ; breast, abdomen and posterior femora pitch-black ; rest of legs red-yellow.　Punctuation rather stronger than in *P. hyoscyami.*　L. 1¾ l.　Scarce.

　　　　　　　　　　　　　　　　P. cyanoptera, Ill.

　　B B. Elytra reddish-brown.

Oblong-ovate.　Head and thorax brassy-green ; elytra reddish-brown, shiny ; antennae red-yellow ; legs brassy-green, posterior femora brassy, pale at base.　Narrower than *P. hyoscyami*; smaller than *P. chrysocephala*, elytra with more finely and closely punctured striae.　L. 1½ l.　Scarce.

　　　　　　　　　　　　　P. luridipennis, Kuts.

　　2. Under-side pale red-yellow.

Ovate. Reddish-yellow; elytra, tibiae and tarsi paler. Forehead strongly punctured; punctuation of thorax finer and closer than in *P. hyoscyami*; elytra with regular rows of punctures, interstices finely punctured; apical process of posterior tibiae broad and hollowed out toward end, with a fine, pointed tooth in middle. L. 1¾—2¼ l. Rather common. *P. marcida*, Ill.

II. Thorax strongly punctured.

Oblong-ovate. Upper-side bronze-green; under-side bronze-black; base of antennae and legs red-yellow, posterior femora black. Interstices on elytra finely punctured. L. 1—1½ l. Common. *P. cupronitens*, Forst.

B. Forehead with a straight furrow between eyes.

Oblong-ovate. Bronze or black; front pairs of femora brown, posterior femora bronze, tibiae and tarsi red-yellow. Thorax very finely punctured, a little more strongly at sides; elytra with rows of rather shallow punctures, interstices finely wrinkled. L. 1¼ l. Scarce.
P. instabilis, Foudr.

C. Forehead with two diagonal furrows, crossing X-like in middle.

a. Thorax green or bronze; suture of elytra not black.

Oblong-ovate. Bronze-green above; apex of elytra reddish; under-side black; antennae and legs yellow-red, former brownish at apex, latter with posterior femora bronze and base of front pairs of femora often brown. Thorax distinctly punctured; elytra with rows of strong punctures, interstices distinctly punctured and wrinkled. L. 1—1¼ l. Not very common. *P. attenuata*, E.H.

b. Thorax pale red-yellow; suture of elytra black.

Oblong-ovate, not very convex. Head, suture of elytra, under-side and posterior femora black; thorax and rest of elytra, whole of antennae and rest of legs pale red-yellow. Thorax distinctly and closely punctured, often with an impression on each side at base; elytra with regular rows of deep punctures, interstices finely punctured. Whole of upper-side sometimes yellow. L. 1—1¼ l. Common.
P. affinis, Payk.

D. Forehead with a round depression between eyes.

a. Posterior femora yellow-red.

Rather narrower than *P. affinis* and less convex, rows of punctures on elytra placed nearer each other, thorax longer and much less strongly punctured. Lighter or darker yellow-red, under-side usually darker. Thorax indistinctly punctured, with a small impression on each side of base; rows of punctures on elytra enfeebled at apex, interstices very indistinctly punctured. L. 1¼ l. Not very common.

P. luteola, Müll.

b. Posterior femora bronzed-pitch-brown.

Ovate. Pitch-brown, upper-side with bronze reflection; antennae reddish; legs (except posterior femora) yellow-red. Thorax very finely and not very closely punctured, with a slight impression on each side at base; rows of punctures on elytra not very straight at base, interstices with a few wrinkles and very fine punctures. L. 1¼—1½ l. Common.

P. picina, Marsh.

Apteropeda.

A. Interstices on elytra punctured.

Nearly hemispherical, but slightly narrowed before and behind. Upper-side bronze, blue or green; under-side bronze-black; antennae and legs yellow-red, former brownish at apex, latter with front pairs of femora more or less brown and posterior femora bronze or blue. Thorax rather closely punctured; interstices on elytra flat, finely punctured; posterior tibiae toothed. L. 1¼—1½ l. Common.

A. graminis, Panz.

B. Interstices on elytra almost impunctate.

a. Thorax tolerably closely punctured.

Similar in form to *A. graminis*. Pitch-black; antennae and legs reddish; femora of posterior legs bronze-black; their tibiae brownish, toothed; antennae thicker than in *A. graminis*, punctuation of thorax more diffuse and a little stronger at sides; interstices on elytra tolerably convex, with a few very fine punctures. L. 1½ l. Scarce.

A. globosa, Panz.

b. Thorax scarcely punctured.

Similar in form to *A. graminis*. Bluish-black; under-side black; antennae and legs yellow-red, posterior femora pitch-black. Antennae thicker than in *A. graminis*; interstices on elytra rather convex; posterior tibiae not toothed, but fringed at apex. L. 1¼—1½ l. Scarce.

A. splendida, All.

Mniophila.

Upper-side dark-bronze; under-side pitch-brown; antennae and legs yellow-red, posterior femora often darker. Elytra pointed at apex, with rows of very fine punctures, somewhat indistinct on disc and confused at apex. L. ½ l. Rather common. *M. muscorum*, E. H.

Sphaeroderma.

A. Thorax very indistinctly punctured, body hemispherical.

Yellow-red; eyes black. Sides of thorax rounded; punctuation of elytra a little more distinct than that of thorax, here and there in rows; tibiae (especially intermediate pair) curved. L. 1½ l. Common.

S. testacea, Fab.

B. Thorax distinctly punctured; body oblong.

Similar to *S. testacea*, but less convex, sides of thorax longer; punctuation stronger; femora larger, tibiae almost straight. L. 1½—1¾ l. Common. *S. cardui*, Gyll.

CASSIDIDAE.

Antennae placed near each other on forehead; eyes oval; thorax at least twice as broad as long. *Cassida*, Lin.

Cassida.

A. Elytra punctured at least partly in rows.

　a. Elytra without regular raised lines.

　　I. Side margin of thorax and elytra flattened or slightly raised.

　　　1. Elytra red or green, with black spots or lines.

　　　　A A. Side margin of thorax and elytra flattened.

Rounded oval. Upper-side red-brown or green, elytra with black spots; under-side and legs black. Posterior angles of thorax pointed, or at least right angles; elytra with regular rows of punctures. L. 3—3½ l. Moderately common. *C. murraea*, Lin.

 B B. Side margin of thorax and elytra slightly raised.

Upper-side red ; thorax with black spots ; elytra with (together) three longitudinal black streaks, variable in extent; under-side and legs black. Thorax punctured at margins; elytra with fine, regular rows of punctures. L. 2¼ l. Not common. *C. vittata*, Fab.

 2. Elytra green, with base and sometimes suture red-brown.

 A A. Elytra with distinct rows of punctures at suture and shoulders only.

Broad oval. Upper-side green; elytra red-brown at base only; under-side black ; legs greenish, femora black toward apex. Thorax closely and deeply punctured, posterior angles distinct. L. 2¾—3½ l. Common.

 C. viridis, Lin.

 B B. Elytra with distinct rows of punctures throughout.

 a a. Elytra with suture broadly red-brown.

Oval. Upper-side green; elytra as above ; under-side and legs black. Posterior angles of thorax distinct. L. 2½ —3 l. Moderately common. *C. vibex*, Lin.

 b b. Elytra with suture green or only narrowly red-brown.

 A a. Posterior angles of thorax somewhat pointed.

Oval. Upper-side green; elytra with a triangular spot at base and the suture narrowly red-brown; under-side black; legs greenish. Elytra with rows of coarse punctures and some slightly raised longitudinal lines. L. 2—2¼ l. Not uncommon. *C. sanguinolenta*, Fab.

 B b. Posterior angles of thorax pointed.

Rounded oval. Upper-side green ; elytra with base and sometimes suture (narrowly) red ; under-side black ; legs green or yellow. Elytra with rows of punctures and some slightly raised longitudinal lines. L. 2¼—2¾ l. Near Dumfries. *C. chloris*, Suffr.

_II._ Side margin of thorax and elytra directed down-
ward.

   _1._ Posterior angles of thorax pointed.

Elliptic. Upper-side green ; each elytron with a broad
greenish-golden streak ; under-side black, margins of abdomen broadly greenish, legs green. Thorax finely punctured
on disc, more strongly toward margins ; elytra with regular
rows of punctures. L. $2\frac{1}{2}$—3 l. Moderately common.

C. oblonga, Ill.

   _2._ Posterior angles of thorax rounded.

Elliptic. Upper-side pale greenish-yellow ; elytra with
a golden streak on each ; under-side black, margins of abdomen (narrowly) and legs yellow, base of femora black.
Elytra with regular rows of punctures. L. 2—$2\frac{1}{2}$ l. Moderately common. C. nobilis, Lin.

   _b._ Elytra with regular raised interstices.

     _I._ Upper-side rust-brown.

       _1._ Upper-side spotted irregularly with black ;
elytra with regular rows of coarse punctures.

Oval. Under-side black ; head, margins of abdomen
broadly and legs (except base of femora) rust-brown. Posterior angles of thorax rounded. L. $2\frac{1}{4}$—$2\frac{1}{2}$ l. Not common. C. nebulosa, Lin.

       _2._ Upper-side not spotted with black ; elytra with
irregular rows of punctures.

Oval. Under-side black ; legs red. Posterior angles of
thorax rounded ; alternate interstices on elytra more raised.
L. 2—$2\frac{1}{2}$ l. Rare. C. ferruginea, Fab.

     _II._ Upper-side pale greenish-yellow.

Broad elliptic. Under-side black ; head and legs yellow.
Posterior angles of thorax rounded ; elytra with regular
rows of punctures. L. 2—$2\frac{1}{2}$ l. Common.

C. obsoleta, Ill.

_B._ Elytra confusedly punctured throughout.

   _a._ Body oval ; length $3\frac{1}{2}$—$4\frac{1}{2}$ lines.

Upper-side green ; under-side black, margins of abdomen
and legs reddish-yellow. Posterior angles of thorax rounded ;
elytra rather closely punctured, interstices very finely

wrinkled, dull, shoulders rather broader than thorax, very prominent. Common. *C. equestris*, Fab.

b. Body almost circular ; length 2 lines.

Convex. Upper-side green or yellowish-green ; under-side (except abdomen) black ; abdomen and legs yellow. Posterior angles of thorax right angles, but rounded at apex ; elytra deeply punctured, interstices smooth, shiny, side margin divided from disc by a row of deep punctures, shoulders not broader than thorax. Not common.

C. hemisphaerica, Herbst.

LONGICORNIA.

A. Last joint of palpi pointed at apex ; anterior tibiae with an oblique groove on inner-side.

Ligula leathery or horny, rarely membranous, without paraglossae ; maxillae with two lobes ; labrum separate, horizontal ; head usually vertical in front, forehead form-ing a right, or acute angle with vertex ; antennae inserted far from mandibles, in an emargination of eyes ; pronotum scarcely ever separate from sides of prothorax ; mesonotum with organs of stridulation ; socket holes of anterior coxae closed behind. *Lamiidae.*

B. Last joint of palpi truncate at apex ; anterior tibiae without oblique groove on inner-side.

a. Ligula generally membranous ; pronotum very rarely separate from sides of prothorax ; anterior coxae very variable.

Maxillae with two lobes ; labrum not soldered to clypeus, never vertical compared with latter ; mesonotum nearly always with organs of stridulation. *Cerambycidae.*

b. Ligula horny ; pronotum separate from sides of pro-thorax ; anterior coxae strongly transverse.

Inner lobe of maxillae very rarely distinct ; labrum soldered to clypeus ; antennae usually inserted near base of mandibles before eyes ; mesonotum without organs of stridulation ; socket holes of anterior coxae nearly always widely open behind. *Prionidae.*

LAMIIDAE.

A. Tarsal claws simple.

a. First joint of antennae with cicatrice at apex.

I. Thorax with lateral spines.

1. Antennae shorter than, or at most as long as body.

Cicatrice of first joint of antennae large; metasternum slightly elongate; femora about equal in thickness throughout; anterior legs of male not longer than the others.

Lamia, Fab.

2. Antennae much longer than body.

Cicatrice of first joint of antennae large; thorax with two transverse furrows; metasternum elongate; femora about equal in thickness throughout; anterior legs of male longer than the others. *Monohammus,* Serv.

II. Thorax without lateral spines.

Antennae much longer than body, ridge bounding cicatrice of first joint interrupted; eyes nearly divided; thorax without transverse furrows; metasternum moderately elongate; femora gradually clubbed; legs about equal.

Mesosa, Serv.

b. First joint of antennae without apical cicatrice.

I. Thorax with lateral spines or tubercles.

1. Antennae not or scarcely fringed beneath; intermediate tibiae with furrow.

A A. Antennae with a few short hairs beneath; lateral tubercle of thorax placed immediately behind middle.

Antennae of male from three to five times, those of female about twice as long as body; elytra flattened on disc, entire at apex; last abdominal segment emarginate in male, much prolonged in female, femora gradually clubbed.

Acanthocinus, Steph.

B B. Antennae without hairs; lateral spine on thorax placed near base.

Antennae much longer than body; eyes more or less approximated above; elytra somewhat convex.

Liopus, Serv.

2. Antennae with rather long fringe beneath; intermediate tibiae entire.

Antennae a little longer than body ; elytra with lateral ridge at base, gradually narrowed behind, apex truncate and often spined at outer angles, each with three abbreviated ridges ; femora strongly clubbed.   *Pogonocherus*, Serv.

II. Thorax without lateral spines or tubercles.

1. Antennae with twelve joints.

Antennae longer than body, more or less fringed beneath ; elytra moderately long, moderately convex, narrowed and obtusely rounded at apex ; femora nearly linear, posterior pair reaching a little beyond second abdominal segment ; body elongate.   *Agapanthia*, Serv.

2. Antennae with eleven joints.

Antennae at most a little longer than body, feebly or scarcely fringed beneath ; elytra more or less elongate, flat or very little convex, parallel-sided or gradually narrowed behind ; femora nearly linear, posterior pair reaching to apex of fourth abdominal segment ; body elongate.

Saperda, Fab.

B. Tarsal claws split, toothed or appendiculate.

a. Eyes not divided.

I. Posterior femora reaching beyond second abdominal segment.

1. First tarsal joint shorter than second and third together.

Antennae as long as, or a little longer than body, feebly fringed beneath ; thorax without lateral tubercles ; elytra usually moderately elongate, flattened on disc, nearly parallel or a little narrowed behind ; femora gradually clubbed, posterior pair reaching more or less beyond second abdominal segment ; intermediate tibiae with furrow.   *Phytoecia*, Mulo.

2. First tarsal joint a little longer than second and third together.

Antennae a little longer than body, feebly fringed beneath, third joint very elongate ; thorax without lateral tubercles ; elytra nearly flat, elongate, parallel-sided ; posterior legs much longer than the others ; femora gradually and feebly thickened, posterior pair reaching to apex of fourth abdominal segment ; intermediate tibiae with very indistinct furrow, often with none.

Stenostola, Muls.

II. Posterior femora reaching at most to apex of second abdominal segment.

Antennae usually rather shorter than body ; thorax without lateral tubercles ; elytra very elongate, flat; legs short, equal, femora gradually clubbed ; intermediate tibiae with furrow. *Oberea*, Muls.

b. Eyes divided.

Antennae shorter than body, fringed ; thorax without lateral tubercles, with a basal furrow ; elytra moderately elongate, parallel-sided, rather flat ; legs short, femora gradually clubbed, posterior pair scarcely reaching beyond apex of second abdominal segment. *Polyopsia*, Muls.

Lamia.

Convex. Black, dull ; with very fine, close-lying brown pubescence. Thorax punctured in wrinkles ; elytra granulate. L. 10—14 l. Scarce. *L. textor*, Lin.

Monohammus.

A. Scutellum entirely covered with yellowish pubescence.

Black, with brownish-bronze reflection ; elytra with indistinct spots of grayish-yellow hairs (more apparent in female) arranged somewhat in bands. Elytra punctured in wrinkles at base and toward apex ; first joint of posterior tarsi longer than third. L. 10—13 l. Rare.

M. sartor, Fab.

B. Scutellum covered with pale yellowish pubescence, except on a central longitudinal line.

Similar to *M. sartor*. Elytra punctured more strongly at base than at apex ; first joint of posterior tarsi scarcely longer than third. L. 8—10 l. Rare. *M. sutor*, Lin.

Mesosa.

Black ; thickly covered with brownish-gray pubescence, with black lines on head and thorax and with a more or less distinct whitish interrupted band, bordered with black, in middle of elytra. L. 4—5 l. Not uncommon.

M. nubila, Ol.

Acanthocinus.

Brown; with thick gray pubescence; elytra with two irregular brownish bands and with lines of black points; first joint of antennae black along outer-side. L. 6—8 l. Not common.   *A. aedilis*, Lin.

Liopus.

Black; closely covered with gray pubescence; thorax and elytra with cloudy spots, forming on latter a brown band behind middle; joints of antennae, femora and tarsal joints darker at apex than at base; tibiae with a ring of light pubescence near base. L. 3—4 l. Rather common.

L. nebulosus, Lin.

Pogonocherus.

A. Outer apical angle of elytra not produced.

Brown; pubescence gray and brownish; elytra with a broad oblique band of white hairs behind base and with three raised longitudinal lines, the innermost bearing three spots of black hairs toward apex. L. 2½—3 l. Not common.

P. fasciculatus, De G.

B. Outer apical angle of elytra produced into a spine.

a. Sutural apical angle of elytra produced into a small spine.

Brown; pubescence not very thick, gray and brownish; elytra with a broad band of white hairs before middle and with three raised longitudinal lines, the innermost bearing three spots of black hairs toward apex. L. 3 l. Moderately common.   *P. hispidus*, Lin.

b. Sutural apical angle of elytra not produced.

Brown; pubescence not very thick, yellowish-gray; elytra with a broad band of yellowish-white hairs behind base and with three longitudinal raised lines, united behind, the innermost bearing two spots of white hairs. L. 2½— 3 l. Not uncommon.   *P. dentatus*, Fourc.

Agapanthia.

A. Length 6—8 lines.

Black; with rather short, yellow pubescence (on elytra in spots) and also with long, erect black hairs; central line and sides of thorax and the scutellum closely covered with yellow hairs; antennae reddish-yellow, with white hairs, first joint and apex of each following joint black. Moderately common. *A. lineatocollis,* Don.

B. Length 4½—5 lines.

Purple or dark violet; a line at sides of thorax, scutellum and sides of breast closely covered with white hairs. Punctuation of elytra much coarser than that of head and thorax. (? Brit.) *A. micans,* Payk.

Saperda.

A. Elytra considerably narrowed toward apex.

Black; thickly covered with yellowish pubescence, sprinkled with black points. Elytra pointed at apex. L. 10—12 l. Moderately common. *S. carcharias,* Lin.

B. Elytra not much narrowed toward apex.

a. Elytra flattish, black, with suture (dilated in six places), and some spots at outer margin covered with greenish-yellow pubescence.

Head and thorax black, covered (except a triangular spot on former and part of disc of latter) with greenish-yellow pubescence. L. 6—8 l. Scarce. *S. scalaris,* Lin.

b. Elytra rather convex, black, with yellowish-gray pubescence, four spots on each more thickly covered with yellowish pubescence.

Head and thorax black, with yellowish-gray pubescence and latter with three lines of yellowish hairs. L. 5—6 l. Rather common. *S. populnea,* Lin.

Phytoecia.

Black; pubescence grayish; central line of thorax and the scutellum covered with whitish hairs; anterior legs (except base of femora) reddish-yellow. L. 4 l. Not common. *P. cylindrica,* Lin.

Stenostola.

Upper-side black, with bluish or greenish reflection; under-side and legs black; pubescence grayish; a line on each side of thorax, sides of breast and usually the scutellum covered with whitish hairs. L. 4½—5 l. Scarce.

S. ferrea, Schr.

Oberea.

Reddish-yellow; head, two points on thorax, elytra and antennae black; elytra covered with close, short, silver-gray pubescence. L. 7—8 l. Not common.

O. oculata, Lin.

Polyopsia.

Black; elytra (except apex) and legs (except hinder femora) yellow-brown. L. 2 l. Common.

P. praeusta, Lin.

CERAMBYCIDAE.

A. Head constricted into a neck behind; anterior coxae prominent, reaching beyond level of prosternal process (except in *Rhagium*).

a. Antennae inserted in emargination of eyes.

Antennae of male considerably longer, of female scarcely longer than body, those of male more or less distinctly twelve-jointed; eyes strongly emarginate, finely granulate; thorax elongate, with slight lateral tubercles; elytra only slightly longer than together broad, separately rounded at apex; wings exposed; body narrow.   *Molorchus*, Fab.

b. Antennae inserted outside eyes; latter entire or nearly so, finely granulate.

1. Anterior coxae not or scarcely prominent.

Antennae not reaching beyond middle of elytra, generally rather stout; head suddenly constricted behind (more or less far from eyes); thorax as long as broad, with conical lateral tubercles; elytra moderately elongate, rather flat, somewhat narrowed behind, apex rounded.

Rhagium, Fab.

II. Anterior coxae strongly prominent.

1. Fourth joint of antennae much shorter than fifth.

Antennae as long or nearly as long as body; head gradually narrowed behind; thorax about as long as broad, with obtuse lateral tubercles; elytra elongate, somewhat flat, gradually narrowed behind, apex more or less truncate, sometimes spined. *Toxotus*, Serv.

2. Fourth joint of antennae nearly as long as fifth.

A A. Head gradually narrowed behind.

Antennae as long or nearly as long as body; thorax narrowed in front, sometimes with lateral tubercles.
Pachyta, Serv.

B B. Head abruptly narrowed behind.

a a. Elytra narrowed from base to apex.

A a. Posterior angles of thorax produced spinelike.

Antennae as long or nearly as long as body; thorax usually longer than broad, narrowed in front, sides rounded, without lateral tubercles. *Strangalia*, Serv.

B b. Posterior angles of thorax obtuse.

Antennae scarcely as long as body; thorax as long as broad, narrowed in front, sides rounded, without lateral tubercles. *Leptura*, Lin.

b b. Elytra not, or only slightly narrowed behind.

A a. Posterior angles of thorax obtuse.

Antennae as long or nearly as long as body; thorax rounded at sides, without lateral tubercles.
Anoplodera, Muls.

B b. Posterior angles of thorax spined.

Antennae scarcely as long as, or shorter than body; thorax rounded at sides, without lateral tubercles.
Grammoptera, Serv.

B. Head not constricted into a neck behind; anterior coxae not or only slightly prominent.

a. Thorax without any, or with only very feeble lateral tubercles.

I. Thorax longer than broad ; eyes strongly granulate.

1. First abdominal segment of normal length.

Maxillary palpi at least three times as long as labial, last joint of all slightly dilated at apex ; antennae as long as, or slightly longer than body ; elytra elongate, parallel-sided, rounded at apex.   *Gracilia*, Serv.

2. First abdominal segment nearly as long as all the rest together.

Palpi short, maxillary a little longer than labial, last joint of all nearly threadlike, obtuse at apex ; antennae about as long as body ; elytra double as long as head and thorax, parallel-sided, rounded at apex.   *Obrium*, Serv.

II. Thorax as broad as, or broader than long; eyes finely granulate.

1. Anterior coxae globular.

Antennae rather stout, shorter than body ; eyes somewhat broadly and strongly emarginate ; thorax very convex ; elytra moderately elongate, more or less parallel-sided, apex obliquely truncate ; femora gradually clubbed.   *Clytus*, Fab.

2. Anterior coxae strongly angular on outer-side.

A A. Eyes not strongly emarginate ; femora thickened in middle.

Antennae shorter than half body, third joint not much longer than fourth ; thorax transverse, moderately convex ; elytra rather short, moderately convex, parallel-sided, rounded at apex.   *Asemum*, Esch.

B B. Eyes strongly emarginate ; femora thickened at apex.

a a. Anterior coxae placed apart.

Antennae scarcely reaching middle of elytra, third joint double as long as fourth ; thorax transverse, rather flat, sides rounded, disc with two shining callosities ; elytra rather flat, moderately elongate, parallel-sided, apex rounded.   *Hylotrupes*, Serv.

b b. Anterior coxae placed near each other.

Antennae a little shorter than body, third joint a little longer than fourth ; thorax transverse, flat, sides strongly rounded; elytra flat, parallel-sided or dilated behind, apex rounded. *Callidium*, Fab.

b. Thorax with strong lateral spines.

Antennae as long as or somewhat longer than body; thorax broader than long; elytra flat, parallel-sided, flexible; anterior coxae globular. *Aromia*, Serv.

Molorchus.

A. Elytra brown, with an oblique white line on each.

Black, hairy; elytra (except white line), antennae and legs (except apex of femora) brown. L. 4 l. Rare.

M. minor, Lin.

B. Elytra unicolorous brown.

Black, hairy; elytra, antennae and legs (often except apex of femora) brown. L. 3½ l. Not common.

M. umbellatarum, Lin.

Rhagium.

A. Antennae entirely black, strongly pubescent.

a. Elytra black, irregularly strewn with spots of yellowish-gray pubescence, with two red-yellow bands, separated at outer margin by a smooth black spot.

Head, thorax, under-side and legs black, with yellowish-gray pubescence. Elytra with two raised lines. L. 7—9 l. Common. *R. inquisitor*, Fab.

b. Elytra pale yellow-brown, covered with gray pubescence, with two bands and some markings black.

Head, thorax, under-side and legs black, latter reddish beneath, with gray pubescence. Elytra with three raised lines. L. 6—7 l. Moderately common.

R. indagator, Lin.

B. Antennae red-brown at apex, feebly pubescent.

Black, with feeble gray pubescence; elytra red-brown at sides and apex, each with two oblique yellow spots and three or four raised lines; femora and tibiae red-brown at base. L. 7—8 l. Common. *R. bifasciatum*, Fab.

Toxotus.

Head and thorax black; elytra wholly or partly black, or entirely yellow-brown; under-side black, part of abdomen sometimes brown; part of antennae and of legs yellow-brown; pubescence gray or yellowish. L. 6—10 l. Common.   *T. meridianus*, Lin.

Pachyta.

A. Thorax red.

Head black; elytra violet or dark blue; breast black; abdomen red. L. 3—3½ l. Rather common.

P. collaris, Lin.

B. Thorax black.

Head black; elytra dull pale yellow, with black apex and four black spots on each (three near side margin and one on disc near base); under-side black. L. 4—5 l. Not uncommon.   *P. octomaculata*, Fab.

Strangalia.

A. Elytra yellow, with black markings, or black, with yellow markings.

　a. Base of femora black.

　　I. Margins of thorax closely covered with golden-yellow hairs, disc with black hairs.

Black; elytra with four yellow bands, interrupted at suture; legs (except base of femora) red-yellow; antennae of female red-yellow. Margins of abdominal segments fringed with yellow hairs. L. 7—8 l. Rare.

S. aurulenta, Fab.

　　II. Thorax with feeble yellowish-gray pubescence.

Black; elytra with four yellow bands, nearly reaching suture; apex of antennae of female red-yellow. L. 7 l. Scarce.   *S. quadrifasciata*, Lin.

　b. Base of femora yellow.

　　I. Sides of thorax angular in middle.

Black; part of elytra (markings variable), base of joints of antennae and the legs (except apex of posterior femora) yellow; abdomen yellow at base in female. Posterior tibiae of male with two teeth on inner-side. L. 6—7 l. Rather common. *S. armata*, Herbst.

II. Sides of thorax rounded in middle.

Black; elytra with four red-yellow bands; part of base of abdomen and the legs (except apex of femora and tibiae on hinder pairs) reddish-yellow; apex of antennae lighter than base. L. 5—6 l. Rare. *S. attenuata*, Lin.

B. Elytra unicolorous black or reddish-yellow.

a. Head and thorax reddish-yellow.

Reddish-yellow; elytra black or reddish-yellow; breast black; antennae black toward apex. Thorax as long as broad. L. 5 l. Rare. *S. revestita*, Lin.

b. Head and thorax black.

Black, shiny; part of abdomen reddish; legs pale yellowish. Thorax longer than broad, much narrowed in front. L. 3½—4 l. Not uncommon. *S. nigra*, Lin.

C. Elytra reddish or yellow-brown, with suture and apex black.

Black (except elytra); head and thorax dull. L. 4—4½ l. Common. *S. melanura*, Lin.

Leptura.

A. Upper-side closely, under-side very closely covered with yellowish-green pubescence.

Body black; joints of antennae yellow at base. L. 6—8 l. Rare. *L. virens*, Lin.

B. Pubescence gray, not close.

a. Legs red.

Black; mouth, elytra, part of abdomen, antennae of female and legs rust-red. L. 7 l. Rare. *L. rufa*, Brul.

b. Legs black (anterior tibiae sometimes brown).

I. Elytra black.

Black, dull; scutellum closely covered with yellowish pubescence. Punctuation strong, finer toward apex of elytra. L. 7 l. Scarce. *L. scutellata*, Fab.

II. Elytra either yellow-brown, with apex black, or entirely bright-red.

1. Anal segment distinctly emarginate; posterior femora rather thick and not much longer than those of front pairs of legs.

Black; elytra (except apex) light yellow-brown. Thorax closely punctured. L. 5 l. Scarce.   *L. fulva*, De G.

2. Anal segment not distinctly emarginate; posterior femora long and thin.

Black; elytra of male yellow-brown, with apex and part of outer margin black, those of female bright-red. Thorax very closely punctured. L. 4½ l. Rare.

L. sanguinolenta, Lin.

III. Elytra entirely yellow-brown.

Black (except elytra). Thorax strongly, not closely punctured. L. 3 l. Common.   *L. livida*, Fab.

Anoplodera.

Black; each elytron with three reddish-yellow spots, the two hinder ones sometimes united. Thorax almost longer than broad, very closely punctured. L. 4—5 l. Not common.   *A. sexguttata*, Fab.

Grammoptera.

A. Elytra yellow-brown, with suture and apex blackish.

Head, thorax and under-side black (anal segment often reddish); elytra as above; legs yellow-red, apex of femora sometimes dark; pubescence yellowish. L. 2½—3 l. Common.   *G. tabacicolor*, De G.

B. Elytra black.

  a. Femora and tibiae more or less black.

    I. Antennae entirely black; length 3½—4 lines.

Black; apex of abdomen and base of femora reddish; pubescence pale-yellow. Not uncommon.

G. analis, Panz.

    II. Antennae with first two joints red, the others red at base and black at apex of each; length 2½—2¾ lines.

Black; anterior legs and base of hinder femora reddish-yellow ; pubescence pale yellow. Common.

G. ruficornis, Fab.

b. Femora and tibiae entirely reddish-yellow.

Black ; antennae brownish, first two joints lighter ; pubescence bright yellow, scanty on head, thorax and apex of elytra. L. 3½—4 l. Not uncommon.

G. praeusta, Fab.

Gracilia.

Lighter or darker brown, dull ; antennae and legs reddish. L. 2—2½ l. Common. *G. pygmaea*, Fab.

Obrium.

Reddish-yellow-brown ; antennae and legs blackish, pubescent. L. 3—4 l. Not common.

O. cantharinum, Lin.

Clytus.

A. Posterior femora gradually thickened toward apex.

a. Thorax much broader than long.

Black, velvety ; thorax with a yellow band at apex and another interrupted one in middle ; scutellum yellow ; elytra with two spots at base of each, three curved bands on disc and an oblique line at sutural angle yellow. L. 5 —7 l. Moderately common. *C. arcuatus*, Lin.

b. Thorax about as long as broad.

Black, velvety ; part of antennae and legs reddish ; thorax with base and apex, elytra with a transverse spot near shoulder, an oblique band (curved toward base) just before middle, another (nearly straight) behind middle and the apex yellow ; scutellum yellow. L. 4½—6 l. Common.

C. arietis, Lin.

B. Posterior femora thin at base, strongly clubbed at apex.

Black ; elytra generally red-brown at base, the apex and three curved lines in middle covered with whitish-gray pubescence. Thorax about as long as broad. L. 4½—6 l. Rather common. *C. mysticus*, Lin.

Asemum.

Black, dull. Upper-side very finely and closely punctured; elytra with striae, interstices more or less raised. L. 6—7½ l. Not very common.   *A. striatum*, Lin.

Hylotrupes.

Pitch-black or brownish; pubescence gray, on thorax close and long, leaving bare only an indistinct central line and a shiny tubercle on each side of disc; elytra sometimes with a more or less apparent band of closer gray pubescence before middle. L. 6—8 l. Scarce.
H. bajulus, Lin.

Callidium.

A. Thorax without tubercles.

a. Elytra unicolorous.

I. Upper-side blue or violet, not pubescent.

Under-side brownish. Sides of thorax rounded. L. 5—6 l. Common.   *C. violaceum*, Lin.

II. Upper-side black, but covered with red pubescence.

Under-side black. Sides of thorax slightly angular. L. 4—5 l. Rare.   *C. sanguineum*, Lin.

b. Elytra with white markings.

Black; antennae, base of elytra and the legs (except apex of femora) rust-red; elytra with two bent bands of white pubescence. Thorax less flat and transverse than in *C. violaceum*, sides rounded. L. 2—2½ l. Common.
C. alni, Lin.

B. Thorax with four smooth tubercles.

Colour variable. Either black, with elytra blue or yellowish-red; or black, with thorax (wholly or partly), part of antennae and legs and apex of abdomen brownish-red, elytra bluish, or yellowish-red, with breast blackish, and with or without head and apex of elytra blackish. Elytra elongate, finely punctured, with a slightly raised line; mesosternum bluntly pointed or rounded between intermediate coxae. L. 5—6½ l. Not common.
C. variabile, Lin.

Aromia.

Bluish-green or coppery-green. Elytra extremely closely and finely wrinkled, with two slightly raised lines. L. 10 —15 l. Common. *A. moschata*, Lin.

PRIONIDAE.

Antennae shorter than body, stout, pectinate or imbricate, rarely sawlike in male ; thorax very transverse.

Prionus, Geoffr.

Prionus.

Pitch-black ; under-side brown ; breast covered with gray pubescence. Thorax with three teeth on each side ; elytra punctured in wrinkles, with three indistinct raised lines ; elytra of male distinctly twelve-jointed, of female eleven jointed. L. 12—18 l. Not common.

P. coriarius, Lin.

HETEROMERA.

A. Socket holes of anterior coxae closed behind, with either anterior coxae not contiguous or tarsal claws comblike.

a. Tarsal claws simple.

Mentum placed in an emargination or carried by a peduncle of submentum ; ligula hidden or not, paraglossae present ; maxillae with inner lobe smaller than outer, often ending in a horny hook; eyes generally emarginate ; antennae inserted under a frontal projection ; coxae not contiguous, anterior pair globular ; sometimes slightly transverse. *Tenebrionidae.*

b. Tarsal claws comblike.

Mentum carried by a peduncle of submentum ; ligula prominent, paraglossae rather indistinct ; maxillary lobes fringed ; eyes nearly always emarginate, sometimes approximated above ; antennae with base free or only slightly covered; anterior coxae globular or slightly transverse, sometimes cylindrical and prominent, in latter case con-tiguous; intermediate pair with trochantina. *Cistelidae.*

B. Socket holes of anterior coxae open behind, or if closed (*Lagriidae*) with anterior coxae contiguous and tarsal claws simple.

　　a. Mentum not carried by a peduncle of submentum.

　　　I. Head not constricted into a neck behind.

　　　　1. Thorax narrower at base than elytra.

Ligula more or less prominent; maxillary lobes fringed; head horizontal or slightly inclined; eyes entire; antennae threadlike, gradually thickened, or ending in a little club, base free; pronotum scarcely ever separate from sides of prothorax; legs short, anterior coxae conico-cylindrical, almost invariably contiguous; penultimate tarsal joint entire, claws simple. 　　　　　　　　　　　*Pythidae.*

　　　　2. Thorax not narrower at base than elytra.

Ligula more or less prominent; maxillary lobes fringed; maxillary palpi large, often sawlike, last joint hatchet-shaped; head bent downward; eyes almost invariably emarginate; antennae nearly always threadlike or slightly thickened toward apex, base free; pronotum separate from sides of prothorax; anterior coxae variable; intermediate coxae with trochantina; penultimate tarsal joint often bilobed, claws nearly always simple. 　　*Melandryidae.*

　　II. Head constricted into a neck behind (rarely absent in *Lagriidae*, where socket holes of anterior coxae are closed behind.)

　　　1. Socket holes of anterior coxae closed behind.

Ligula horny, prominent; maxillary lobes fringed; head prominent; eyes more or less emarginate; base of antennae free; thorax narrower than elytra at base, pronotum not separate from sides of prothorax; anterior coxae prominent, cylindrical or conical, contiguous or nearly so; inter-mediate coxae with trochantina; tarsal claws simple.
　　　　　　　　　　　　　　　　　　Lagriidae.

　　　2. Socket holes of anterior coxae open behind.

　　A A. Posterior coxae contiguous.

Ligula prominent ; maxillary lobes   fringed ; head prominent, bent downward; base of antennae free; thorax generally narrower than elytra at base, pronotum scarcely ever separate from sides of prothorax; anterior coxae prominent, conical or cylindrical, contiguous ; intermediate coxae with trochantina, sometimes rather indistinct; penultimate tarsal joint almost invariably sub-bilobed, claws simple. *Pedilidae.*

B B. Posterior coxae separated by process of first abdominal segment.

Ligula prominent ; maxillary lobes fringed ; head bent downward ; eyes entire ; antennae threadlike or gradually thickened ; thorax narrower at base than elytra, pronotum not separate from sides of prothorax ; anterior coxae cylindrical, prominent, contiguous; intermediate coxae with trochantina ; penultimate tarsal joint nearly always subbilobed, claws simple. *Anthicidae.*

b. Mentum carried by a peduncle of submentum.

I. Head constricted into a neck behind.

1. Tarsal claws simple or toothed, not split into two unequal portions.

A A. Thorax narrower at base than elytra.

Ligula prominent, bilobed ; maxillary lobes horny, fringed ; head moderately inclined ; antennae sawlike or comblike, base free ; pronotum not separate from sides of prothorax ; anterior coxae elongate, nearly cylindrical, prominent, contiguous ; intermediate coxae with trochantina ; posterior coxae placed a little apart ; penultimate tarsal joint somewhat bilobed, claws simple, dilated at base ; body soft. *Pyrochroidae.*

B B. Thorax not narrower at base than elytra.

a a. Pronotum separated from sides of prothorax.

Ligula prominent, membranous, heart-shaped, maxillary lobes membranous, fringed, not soldered at base ; last joint of palpi hatchet-shaped ; mandibles with membranous border on inner-side ; head vertical, short, resting on anterior coxae, neck sunk in thorax ; eyes oval ; antennae threadlike or slightly toothed, base free ; thorax inclined ;

elytra flat, gradually narrowed, pygidium more or less uncovered; legs long, anterior coxae strong, very prominent,
contiguous; intermediate coxae with trochantina; tarsal
claws simple or toothed. *Mordellidae.*

    b b. Pronotum not separated from sides of
    prothorax (at least in front).

Ligula membranous, more or less prominent; maxillary
lobes fringed, soldered at base, inner one sometimes rudimentary or absent; last joint of maxillary palpi not
hatchet-shaped; mandibles without membranous border on
inner-side; head vertical, resting on anterior coxae, neck
sunk in thorax; eyes emarginate or entire; antennae sawlike or comblike; elytra sometimes covering abdomen, at
others abbreviated and gaping, in latter case wings not
folded under them; legs more or less long, anterior coxae
prominent, contiguous; intermediate coxae with or without
trochantina; penultimate tarsal joint entire, claws toothed,
rarely simple; abdomen with from five to eight segments.
                            Rhipiphoridae.

  2. Tarsal claws split into two unequal portions.

Ligula prominent, sinuate or bilobed; maxillary lobes
horny, fringed, inner one sometimes very small; head much
inclined, neck free; eyes emarginate or entire; antennae
variable; thorax narrower at base than elytra, pronotum
not separate from sides; elytra usually flexible, imperfectly embracing body; anterior coxae nearly cylindrical,
very large, contiguous, directed backward; penultimate
tarsal joint nearly always simple, claws divided into two
branches, lower one usually very slender, rarely replaced
by a tooth. *Meloidae.*

  II. Head not constricted into a neck behind.

Ligula prominent, bilobed, lobes divergent and rounded;
outer maxillary lobe longer than inner; mandibles nearly
always bifid at apex, with a membranous border and fringed
on inner-side; antennae long, nearly always threadlike,
base free; thorax narrower than elytra at base, pronotum
not separate from sides; elytra elongate, usually imperfectly embracing abdomen; legs long, anterior coxae elongate, nearly cylindrical, contiguous, prominent; penultimate
tarsal joint nearly always sub-bilobed, claws simple.
                               Oedemeridae.

TENEBRIONIDAE.

A. Under-side of tarsi spiny or bristly, or if with fine hairs (*Heliopathes*), also with metasternum very short and eyes divided.

a. Last joint of maxillary palpi hatchet-shaped.

I. Eyes kidney-shaped; apical spines of tibiae long.

1. Ligula nearly hidden by mentum; intercoxal process broad, quadrangular.

Head more or less prominent, rhomboidal, narrowed behind; labrum prominent; clypeus slightly emarginate in front; inner lobe of maxillae with a horny hook; scutellum generally small; elytra with reflexed margin broad, apex often produced; anterior femora not toothed; anterior tibiae with two rather stout apical spines; hinder pairs of tibiae round; metasternum very short. *Blaps*, Fab.

2. Ligula prominent; intercoxal process narrow, triangular.

Head sunk in thorax; labrum scarcely prominent; clypeus entire; inner lobe of maxillae without hook; thorax transverse; scutellum small; tibiae slightly triangular, with two slender spines, anterior pair finely toothed; metasternum very short. *Crypticus*, Latr.

II. Eyes nearly or quite divided; apical spines of tibiae short.

1. Anterior tarsi of male dilated, hairy beneath; clypeus not strongly emarginate.

Head sunk in thorax to eyes; last joint of labial palpi triangular; penultimate joints of antennae not longer than broad; mandibles split at apex; cheeks much more prominent than eyes; thorax imperfectly or not contiguous to elytra, moderately emarginate in front; anterior tibiae strongly triangular, the others slightly compressed; metasternum very short. *Heliopathes*, Muls.

2. Anterior tarsi of male not dilated, spiny beneath; clypeus rounded in front and strongly triangularly emarginate.

Head sunk in thorax to eyes; last joint of labial palpi ovoid, pointed; penultimate joints of antennae transverse; mandibles split at apex; cheeks much more prominent than eyes ; thorax more or less contiguous to elytra, transverse, deeply emarginate in front, base generally strongly bi-sinuate ; anterior tibiae slightly or moderately dilated, the others rounded ; metasternum generally more or less elongate ; intercoxal process moderately broad, nearly parallel-sided. *Hopatrum*, Fab.

b. Last joint of maxillary palpi not hatchet-shaped.

I. Eyes divided.

Head sunk in thorax to eyes ; last joint of all palpi elongate ; penultimate joints of antennae transverse ; mandibles split at apex ; clypeus rounded in front and triangularly emarginate ; cheeks more prominent than eyes ; thorax strongly transverse, scarcely emarginate in front, base feebly bisinuate ; anterior tibiae strongly triangularly dilated, toothed on outer-side, outer apical angle prominent, the others compressed, with some small spines on outer edge ; metasternum somewhat elongate ; intercoxal process short, narrow, triangular. *Microzoum*, Redt.

II. Eyes kidney-shaped.

Head more or less sunk in thorax ; last joint of palpi very elongate triangular ; antennae longer than head, penultimate joints transverse ; mandibles split at apex ; clypeus entire ; cheeks not more prominent than eyes ; thorax contiguous to elytra, moderately emarginate in front, base truncate ; anterior tibiae triangularly dilated, the others gradually thickened, at least posterior pair with short spines, all with rather long apical spines ; metasternum very short ; intercoxal process triangular, truncate in front. *Phaleria*, Latr.

B. Under-side of tarsi with fine hairs, with either meta-sternum elongate or eyes not divided.

a. Anterior coxae cylindrical.

I. Antennae partly received when at rest into a transverse furrow on head.

1. Eyes completely divided ; sides of thorax finely notched.

Head sunk in thorax to eyes ; last joint of maxillary palpi nearly cylindrical ; mandibles bifid at apex ; cheeks more prominent than eyes ; thorax transverse, strongly and quadrangularly emarginate in front ; tibiae simple, apical spines very short or absent; metasternum elongate; intercoxal process rather narrow, triangular.

Bolitophagus, Ill.

2. Eyes not completely divided; sides of thorax not notched.

Head very short, sunk in thorax to eyes; last joint of maxillary palpi nearly cylindrical; mandibles bifid at apex; cheeks slightly more prominent than eyes; thorax strongly transverse, not or scarcely emarginate in front; tibiae simple, apical spines very short or absent; metasternum elongate; intercoxal process rather narrow triangular.

Heledona, Latr.

II. Head without transverse furrow.

1. Eyes more prominent than cheeks; intermediate coxae with trochantina; apical spines of tibiae indistinct.

A A. First joint of posterior tarsi short.

Head sunk in thorax; last joint of maxillary palpi elongate, slightly depressed and rounded at apex; mandibles bifid at apex; eyes emarginate; tibiae slightly and gradually dilated; metasternum elongate; intercoxal process short, triangular.

Diaperis, Geoffr.

B B. First joint of posterior tarsi elongate.

a a. Intercoxal process broad, nearly parallel-sided.

Head sunk in thorax; last joint of maxillary palpi elongate; mandibles bifid at apex; eyes emarginate; tibiae slightly and gradually dilated; metasternum elongate.

Scaphidema, Redt.

b b. Intercoxal process narrow, triangular.

A a. Last joint of maxillary palpi triangular, nearly equilateral.

Head sunk in thorax; mandibles bifid at apex; eyes emarginate; tibiae slightly and gradually dilated, metasternum elongate.

Platydema, Lap.

B b. Last joint of maxillary palpi elongate triangular.

Head sunk in thorax; mandibles bifid at apex; eyes

Y 2

emarginate ; tibiae linear ; metasternum elongate.
                                    Alphitophagus, Steph.
2. Eyes not more prominent than cheeks ; intermediate coxae without trochantina ; apical spines
of tibiae distinct, short.

A A. Antennae ending in a more or less distinct
club ; head prominent.

Last joint of maxillary palpi long oval, flattened and
obtuse at apex ; clypeus prominent, trapeziform, thin at
sides, slightly emarginate in front ; eyes divided half-way
through, lower portion distinctly larger than upper ; elytra
elongate, parallel-sided ; metasternum elongate ; intercoxal
process not broad, triangular.   *Tribolium*, Mc Leay.

B B. Antennae gradually thickened ; head not
prominent.

a a. Thorax not or only slightly transverse,
with base truncate or nearly so.

A a. Eyes deeply emarginate.

Last joint of maxillary palpi hatchet-shaped, a little
elongate and obliquely truncate ; clypeus of male a little
narrowed and forming a rounded prominence in front, that
of female slightly dilated on sides, widely rounded in front ;
mandibles of male with a more or less long horn, curved
upward, simple and hooked at apex, those of female hidden
beneath clypeus ; eyes nearly entirely divided, lower portion
much larger than upper ; antennae a little shorter than
thorax, third joint longer than the following ones, joints
four to eight obconical, very short, last three a little dilated ;
elytra more or less elongate, nearly parallel-sided ; metasternum elongate ; intercoxal process not broad, triangular.
                                    Gnathocerus, Thunb.

B b. Eyes slightly emarginate.

Clypeus more or less prominent, a little raised at sides,
gradually narrowed and truncate in front ; eyes slightly
emarginate ; third joint of antennae distinctly longer than
fourth ; thorax at least as long as broad, rounded or truncate at base ; elytra elongate, nearly cylindrical, nearly
always incompletely covering pygidium ; tibiae very elongate triangular ; metasternum elongate ; intercoxal process
not broad, triangular.   *Hypophloeus*, Fab.

b b. Thorax distinctly transverse, base strongly bisinuate.

Clypeus short, separated from forehead by a fine flexuous furrow, sides rounded, front sinuate or truncate; eyes partly divided, lower portion much larger than upper; third joint of antennae scarcely longer than fourth; thorax transverse, base rather strongly bisinuate; elytra oblong oval; tibiae triangular; metasternum elongate; intercoxal process not broad, triangular. *Alphitobius*, Steph.

b. Anterior coxae globular.

I. Metasternum elongate.

Head free, rhomboidal; clypeus separated from forehead by a very fine furrow, prominent, gradually narrowed and truncate or sinuate in front; inner lobe of maxillae with a horny hook; last joint of maxillary palpi hatchet-shaped; antennae gradually thickened, third joint more or less elongate; eyes emarginate, lower portion much larger than upper; thorax transverse; semicircularly emarginate at base; elytra elongate, parallel-sided; tibiae rounded, apical spines short but distinct. *Tenebrio*, Lin.

II. Metasternum short.

Head more or less prolonged and narrowed behind; clypeus separated by a furrow from forehead, short, gradually narrowed and widely truncate in front; inner lobe of maxillae without horny hook; last joint of maxillary palpi strongly hatchet-shaped; antennae long and slender, penultimate joints longer than broad, third joint elongate; eyes slightly emarginate; thorax rounded at sides, slightly or scarcely emarginate in front, base truncate or rounded; elytra usually oblong oval; tibiae gradually dilated, apical spines very small. *Helops*, Fab.

Blaps.

A. Fifth and sixth joints of antennae at least half as long again as broad.

a. Thorax distinctly narrowed behind; length of apical process of elytra ¾—1 line.

Black, rather dull above, shiny beneath. Body narrower than in *B. mucronata;* clypeus without smooth central line, more prominent at anterior angles; elytra broadest in middle; metasternum transversely furrowed. Male with a tuft of yellow hairs in middle of apex of first abdominal segment. L. 10 l. Rare.   *B. mortisaga*, Lin.

b. Thorax scarcely narrowed behind; length of apical process of elytra $\frac{2}{5}$—$\frac{2}{3}$ line.

Black, rather dull above, shiny beneath. Clypeus with a more or less distinct smooth central line, scarcely prominent at anterior angles; frontal suture angular in middle; elytra broadest behind middle; metasternum transversely furrowed in middle. L. 9—11 l. Common.

B. mucronata, Latr.

B. Fifth and sixth joints of antennae scarcely longer than broad.

Black, dull above, shiny beneath. Clypeus without smooth central line, scarcely prominent at anterior angles; frontal suture straight in middle; elytra broadest before middle; metasternum with slight traces of a transverse furrow at sides only. Male with a tuft of reddish hairs on first abdominal segment. L. 9—11 l. Rather common.

B. similis, Latr.

Crypticus.

Black, slightly shiny; antennae and legs pitch-brown. Head and thorax finely and very closely punctured, posterior angles of latter rather prominent, rounded; elytra more finely and scantily punctured than thorax, with slight traces of arrangement in rows. L. 2½—3 l. Common.

C. quisquilius, Lin.

Heliopathes.

Moderately convex. Black, shiny. Head and thorax closely and deeply punctured, latter broader than long; elytra with punctured striae, deeper behind than at base, interstices punctured, often wrinkled. L. 3½—4 l. Common.   *H. gibbus*, Fab.

Hopatrum.

Black or gray-black, dull, very closely granulate. Thorax much broader than long, sides slightly rounded, posterior angles prominent, with their apex rounded; elytra

with shallow striae, bearing tubercles, interstices granulate.
L. 3½ l.   Common.   *H. sabulosum*, Lin.

Microzoum.

Black, dull.   Head and thorax very closely punctured,
latter broader than long, somewhat narrowed behind, with
three small impunctate spots and a rather deep impression
on each side behind ; elytra very closely punctured,
coarsely but indistinctly wrinkled ; anterior tibiae with
four or five teeth before the dilatation.   L. 1½ l.   Common.
M. tibiale, Fab.

Phaleria.

Oblong oval, not very convex.   Pale brownish-yellow,
with or without a black spot on each elytron.   Thorax
indistinctly punctured, broader than long, slightly narrowed in front; elytra with slight punctured striae.   L.
3—3½ l.   Rather common.   *P. cadaverina*, Fab.

Bolitophagus.

Black, dull ; sometimes brown.   Thorax punctured in
network, with central furrow, sides nearly parallel in front,
rather strongly sinuate behind, finely notched, posterior
angles acute, prominent ; elytra with punctured striae,
feeble near suture, stronger outward.   L. 3—3¼ l.   Not
uncommon.   *B. reticulatus*, Lin.

Heledona.

Very convex.   Black, brown or red-brown, dull.   Thorax
punctured in network, without central furrow, sides curved
and finely notched ; elytra with notched striae, third and
eighth interstices joined behind.   L. 1½ l.   Rather common.
H. agricola, Herbst.

Diaperis.

Black, shiny ; elytra with a broad toothed band at base,
a narrower one behind middle and the apex orange-yellow.
Forehead concave ; thorax broader than long ; elytra with
feeble punctured striae, interstices nearly flat, scantily
punctured.   L. 3 l.   Rare.   *D. boleti*, Lin.

Scaphidema.

Oval, moderately convex. Upper-side bronze-brown, shiny; head and thorax sometimes lighter; under-side red-brown or brownish, prosternum lighter. Thorax much broader than long, narrowed in front, sides almost straight; elytra with rather feeble punctured striae, interstices punctured. L. 2 l. Not common. *S. aenea*, Payk.

Platydema.

Upper-side dark blue or violet; under-side black; mouth, antennae and legs brown, antennae paler at apex. Forehead with a depression in middle; thorax much broader than long; elytra with punctured striae, interstices flat, punctured. L. 3½—3¾ l. Rare. *P. violacea*, Fab.

Alphitophagus.

Head brown, mouth paler; thorax red, usually with a black spot in front; elytra brown-black, with a transverse band at base, another just behind middle and generally the apex reddish-yellow; under-side yellow-red or red-brown, antennae and legs reddish-yellow. Thorax much broader than long; elytra with punctured striae. L. 1½ l. Moderately common. *A. quadripustulatus*, Steph.

Tribolium.

A. Antennae with a distinct three-jointed club.

Reddish-yellow-brown or red-brown. Thorax broader than long, sides straight behind; elytra with striae, those next suture feeble, interstices very finely punctured. L. 2 l. Rather common. *T. ferrugineum*, Fab.

B. Club of antennae gradually formed by last four or five joints.

Rather larger than *T. ferrugineum*, broader and flatter, with thorax more rounded outwardly in front; anterior tibiae less acutely produced on outer-side; elevated clypeal ridge carried farther back along the eye, which, therefore, looks smaller; elytra more shiny, minute interstitial punctures less regular; antennae with shorter and stouter basal joints, and gradually dilated toward apex. L. $1\frac{3}{4}$—2 l. Scarce.   *T. confusum*, Duv.

Gnathocerus.

Somewhat elongate, parallel-sided, only slightly convex. Red-brown. Thorax broader than long, sides straight behind; elytra with rows of punctures, fourth and fifth shorter. Mandibles of male very large and prominent. L. $1\frac{1}{2}$—2 l. Common.   *G. cornutus*, Fab.

Hypophloeus.

A. Reflexed margin of elytra prolonged to sutural angle ; elytra covering or nearly covering pygidium.

Rust-red, slightly shiny. Thorax as long as broad ; elytra not much more than twice as long as together broad, punctured in rows throughout. L. $1\frac{1}{4}$—$1\frac{1}{2}$ l. Not common.   *H. depressus*, Fab.

B. Reflexed margin of elytra not prolonged to sutural angle ; elytra not covering pygidium.

 a. Elytra with rows of punctures (except at apex), unicolorous.

Dark reddish-brown, shiny ; antennae and legs lighter. Thorax longer than broad ; elytra nearly three times as long as together broad. L. $2\frac{1}{2}$—3 l. Not uncommon.   *H. castaneus*, Fab.

 b. Elytra confusedly punctured, bicolorous.

Rust-red ; elytra black from before middle to apex. Thorax slightly longer than broad. L. $1\frac{1}{2}$—$1\frac{3}{4}$ l. Common.   *H. bicolor*, Ol.

Alphitobius.

A. Basal border of thorax interrupted in middle.

Upper-side pitch-black, shiny ; under-side brown or rust-red ; antennae and legs brownish. Thorax broadest near base, punctuation stronger at sides than on disc. L. $2\frac{1}{2}$—$2\frac{3}{4}$ l. Rather common.   *A. diaperinus*, Panz.

B. Basal border of thorax not interrupted in middle.

Upper-side pitch-black, slightly shiny ; under-side brown ; apex of antennae and legs red-brown. Thorax broadest

slightly behind middle, punctuation nearly uniform. L.
2½ l. Rather common.   *A. piceus*, Ol.

Tenebrio.

A. Upper-side dull; third joint of antennae nearly as
long as fourth and fifth together, apical joint transverse
oval.

Black; abdomen sometimes brown. Thorax with a
narrow transverse elevation (often indistinct), bounded by
a furrow at each end, before middle of base; interstices on
elytra very slightly convex, each with a row of small
granules; femora very slightly thickened in middle, anterior pair with a rudimentary tooth beneath. L. 6½—8 l.
Rather common.   *T. obscurus*, Fab.

B. Upper-side shiny; third joint of antennae distinctly
shorter than fourth and fifth together, apical joint rather
longer than broad.

Upper-side black or black-brown, under-side rather
lighter. Thorax with few or no traces of transverse
elevation before base; interstices on elytra somewhat convex; femora distinctly thickened in middle. L. 7 l.
Common.   *T. molitor*, Lin.

Helops.

A. Under-side black; raised side margin of elytra almost
interrupted near apex.

Upper-side dark blue or bluish-violet. Thorax a little
broader than long, truncate or nearly so at base, punctured
in wrinkles, sides with a thick raised border; under-side of
femora not fringed. L. 6—8 l. Moderately common.
                                                  H. cocruleus, Lin.

B. Under-side reddish-yellow or red-brown; raised side
margin of elytra uniform.

a. Femora strongly fringed beneath; sides of thorax
sinuate before base.

Reddish-yellow; eyes black. Thorax much broader than
long, truncate at base, rather finely punctured, sides with a
narrow slightly raised border. L. 2½—3½ l. Not uncommon.   *H. pallidus*, Curt.

b. Femora not fringed beneath ; sides of thorax not sinuate before base.

Upper-side brown or black-brown, with bronze reflection ; under-side red-brown. Thorax somewhat broader than long, base bisinuate, sides with a narrow raised border, rather finely punctured. L. 3½—4½ l. Common.

H. striatus, Fourc.

CISTELIDAE.

A. Last joint of maxillary palpi much larger than third ; tarsal claws usually with five or six teeth.

Mandibles slightly bifid at apex ; abdomen with five ventral segments (at least in female).

a. Anterior coxae not contiguous.

Antennae more or less long, threadlike or slightly sawlike, third joint variable in length. *Cistela*, Fab.

b. Anterior coxae contiguous.

Antennae a little shorter than half the body, threadlike, third joint at least as long as fourth ; body more slender than in *Cistela*, eyes more prominent.

Mycetochares, Latr.

B. Last joint of maxillary palpi not much larger than third ; tarsal claws usually with from nine to twelve teeth.

Mandibles entire at apex ; abdomen with six ventral segments.

a. Posterior angles of thorax right angles.

Antennae at least as long as half the body, slender, threadlike or slightly thickened, third joint as long, or nearly so, as fourth ; last joint of maxillary palpi cut very obliquely at apex ; reflexed margin of elytra continued nearly to sutural angle. *Cteniopus*, Sol.

b. Posterior angles of thorax rounded or at least obtuse.

Antennae shorter than half the body, threadlike, apical half a little thickened, third joint longer than fourth ; last joint of maxillary palpi cut less obliquely at apex than in *Cteniopus.* *Omophlus*, Sol.

Cistela.

A. Third joint of antennae shorter than fourth; third joint of posterior tarsi simple.

a. Elytra with punctured striae throughout.

I. Third joint of antennae nearly as large as fourth; posterior angles of thorax nearly right angles, not enclosing shoulders of elytra.

Black, shiny, almost bare; mouth, antennae and legs reddish-yellow; body sometimes pitch-brown or brown, with or without darker head and thorax. L. $3\frac{1}{4}$—$3\frac{3}{4}$ l. Moderately common.   *C. luperus*, Herbst.

II. Third joint of antennae much shorter than fourth; posterior angles of thorax somewhat produced and enclosing shoulders of elytra.

Black, with fine, velvety pubescence; elytra and often also thorax red-yellow. Antennae somewhat sawlike. L. $4\frac{1}{2}$—5 l. Scarce.   *C. ceramboides*, Lin.

b. Elytra with traces of striae near suture only.

Colour variable; usually black, with elytra and legs brown; sometimes brown, with whole or part of thorax, elytra and breast lighter. Upper-side very finely punctured, with fine and rather close pubescence; thorax nearly semicircular, narrowed in front, greatest breadth somewhat behind middle, sides nearly parallel before base. L. $2\frac{1}{2}$ l. Common.   *C. murina*, Lin.

B. Third joint of antennae at least as long as fourth; third joint of posterior tarsi produced beneath fourth joint.

Black, rather shiny; mouth, antennae and legs red-brown. Upper-side finely and not very closely punctured, with semi-erect black hairs; thorax nearly semi-circular, broadest at base; elytra with striae, feeble at base, rather stronger behind. L. 5 l. Not very common.   *C. atra*, Fab.

Mycetochares.

Black; shoulders of elytra, base and apex of antennae, tibiae and tarsi reddish-yellow. Thorax rounded in front, sides nearly parallel from middle to base. L. $2\frac{1}{2}$ l. Not uncommon.   *M. bipustulata*, Ill.

Cteniopus.

Bright sulphur-yellow ; apex of antennae, palpi and tarsi blackish ; eyes black ; head, thorax and under-side of male sometimes darker. Thorax straight in front, broadest before middle, scarcely narrowed behind, much narrowed in front. L. 3—4 l. Common. *C. sulphureus*, Lin.

Omophlus.

Black ; base of antennae, part of maxillary palpi and of tibiae varying from brown to red-yellow ; elytra and tarsi red-yellow. Thorax straight in front, sides slightly curved, greatest breadth in or near middle, with a slight central furrow. L. 4—4½ l. Not common. *O. armeriae*, Curt.

PYTHIDAE.

A. Mandibles prominent.

Last joint of maxillary palpi hatchet-shaped ; intermediate coxae with trochantina ; last abdominal segment not very short ; body flat ; head not prolonged into a rostrum.   *Pytho*, Latr.

B. Mandibles not reaching beyond labrum.

Last joint of maxillary palpi not hatchet-shaped ; intermediate coxae without trochantina ; last abdominal segment very short.

a. Head not prolonged into a rostrum.

I. Antennae gradually thickened.

Mandibles finely toothed on inner-side, bifid at apex ; sides of thorax not toothed.   *Salpingus*, Ill.

II. Antennae ending in a three-jointed club.

Similar to *Salpingus*, but with mandibles not toothed on inner-side and sides of thorax often toothed.

Lissodema, Curt.

b. Head prolonged into a rostrum.

Similar to *Salpingus*, but with mandibles not toothed on inner-side. Antennae inserted on rostrum, latter more or less long, flattened, generally slightly narrowed in middle, more rarely dilated at apex.   *Rhinosimus*, Latr.

Pytho.

Black, shiny, not pubescent; mouth, antennae, tibiae and tarsi rust-red; abdomen wholly or partly yellow; elytra blue, sometimes reddish-yellow-brown at base; margins of thorax occasionally red-yellow. Elytra with deep punctured striae. L. 4½—5½ l. Not uncommon.

P. depressus, Lin.

Salpingus.

A. Labrum as broad again as long.

a. Lateral impressions on thorax before middle as strong as, or stronger than those behind middle; reflexed margin of elytra very narrow from base of abdomen.

I. Sides of thorax with a distinct impression before middle and sometimes a feeble one behind middle; elytra with regular rows of punctures from base,

Metallic-black, sometimes pitchy; tarsi (occasionally legs) and base of antennae red-brown. Thorax scarcely so broad as head; elytra sometimes with a feeble depression before middle. L. 1¼—1½ l. Not common.

S. ater, Payk.

II. Sides of thorax with a distinct impression before middle and another nearly as distinct behind middle; elytra confusedly punctured at base.

Bronze or dark bronze; antennae (except apex) and legs (sometimes except femora) brown-red. Thorax scarcely broader than head; elytra with a transverse depression in middle. L. 1¼—1½ l. Rare.   *S. aeratus,* Muls.

b. Lateral impressions on thorax before middle feebler than those behind middle; reflexed margin of elytra narrow from apex of third ventral segment only.

Dark brown, shiny; base of antennae and greater part of legs paler. Thorax broader than head, with an indistinct impression before middle of side and a stronger one (often connected by a furrow with that on other side) behind middle; elytra confusedly punctured at base, often with a feeble transverse depression before middle. L. 1¼—1½ l. Rather common.   *S. castaneus,* Panz.

B. Labrum less than half as broad again as long.

Bronze-black ; mouth, base of antennae and tarsi rust-red. Thorax rather broader than head, with an impression on each side behind middle and sometimes another indistinct one before middle ; elytra confusedly punctured at base, with a deep depression before middle of each. L. 2 —2¼ l. Scarce. *S. foveolatus*, Ljun.

Lissodema.

A. Elytra unicolorous red-yellow or red-brown.

Red-yellow or red-brown ; legs and often abdomen paler. Thorax broadest nearly in middle ; sides curved behind, with an impression on each side behind middle and sometimes another indistinct one before middle ; elytra with tolerably regular rows of punctures (except at base), with a depression before middle, reflexed margin narrowed from base of abdomen. L. 1½ l. Very rare.

L. Heyana, Curt.

B. Elytra red-brown or blackish, with a reddish-yellow spot at shoulder and often another behind middle.

Head, thorax, base of antennae and prosternum yellowish-red ; legs paler ; under-side brown. Thorax broadest before middle, sides straight behind, with an impression on each side behind middle ; elytra with rather irregular rows of punctures and generally with a depression before middle, reflexed margin narrowed almost from first ventral segment. L. 1—1¼ l. Common. *L. quadripustulata*, Marsh.

Rhinosimus.

A. Thorax red.

a. Vertex of head bronzed or greenish-black.

Yellow-red ; part of forehead, vertex of head and the elytra greenish or bluish-black ; apex of antennae dark. Elytra with rows of punctures, the marginal row not placed in a furrow. L. 1½—2 l. Rather common.

R. ruficollis, Lin.

b. Head entirely yellow-red.

Head, thorax, base of antennae and prosternum yellow-red ; elytra bluish-green ; meso- and metasternum greenish-brown ; abdomen red-brown ; legs yellow. Antennae placed nearer eyes than in *R. ruficollis*; elytra with rows of punctures, the marginal one placed in a furrow in front. L. 1¼—2 l. Rather common.

R. viridipennis, Steph.

B. Thorax bronze-black or brown.

Bronze-black or brown; rostrum, base of antennae and generally under-side yellow-red; legs yellow. Elytra with rows of punctures, the marginal one not or scarcely placed in a furrow. L. 1¼—1½ l. Common.

R. planirostris, Fab.

MELANDRYIDAE.

A. Antennae ending in a large four-jointed club.

Head strongly inclined, scarcely visible from above; maxillary palpi not sawlike, their last joint very slightly hatchet-shaped; apex of mandibles bifid; anterior coxae cylindrical, transverse, separated by a prosternal process reaching to their level; thorax strongly transverse, sides rounded; elytra short, rather convex, parallel-sided.

Tetratoma, Herbst.

B. Antennae threadlike or gradually thickened.

　　a. Anterior coxae not contiguous.

　　　　I. Posterior coxae not oblique; apical spines of tibiae very long.

Head vertical, scarcely or not visible from above; maxillary palpi not sawlike, their last joint hatchet-shaped; apex of mandibles bifid; posterior coxae broad, flat, parallelogrammic; penultimate tarsal joint entire.

Orchesia, Latr.

　　　　II. Posterior coxae oblique; apical spines of tibiae short.

Head vertical, scarcely or not visible from above; maxillary palpi not sawlike, their last joint flattened, parallel-sided, obliquely truncate at apex; apex of mandibles bifid, penultimate tarsal joint entire. *Hallomenus*, Payk.

　　b. Anterior coxae contiguous.

　　　　I. Tarsal claws simple.

　　　　　　1. Antennae eleven-jointed.

A A. Head vertical, scarcely or not visible from above.

a a. Antennae inserted a little above middle of eyes.

A a. Mesosternum not reaching to the half of the intermediate coxae.

Maxillary palpi not or scarcely sawlike, last joint hatchet-shaped, not furrowed on inner edge ; eyes feebly but distinctly emarginate nearly in middle ; front of thorax produced and rounded ; intermediate coxae contiguous behind ; penultimate tarsal joint sub-bilobed.

Anisoxya, Muls.

B b. Mesosternum as long as intermediate coxae.

A 1. Penultimate tarsal joint entire.

a 1. Front of thorax produced and rounded.

Maxillary palpi with fourth joint generally ovoid, depressed, truncate at base and pointed at apex, third joint contiguous to fourth and not prolonged inward, rarely triangular and freer from second ; eyes entire.

Abdera, Steph.

b 1. Front of thorax truncate.

Maxillary palpi large, sawlike, second joint elongate triangular, third transverse triangular, fourth hatchet-shaped, short and prolonged inward ; mandibles bifid at apex ; eyes widely sinuate. *Serropalpus*, Hellen.

B 1. Penultimate tarsal joint sub-bilobed.

Maxillary palpi more or less sawlike, last joint hatchet-shaped, elongate and narrow ; eyes slightly sinuate ; front of thorax produced and rounded ; intermediate coxae not contiguous behind. *Phloeotrya*, Steph.

b b. Antennae inserted nearly level with upper margin of eyes.

A a. Intermediate coxae not contiguous ; third joint of antennae not longer than fourth.

Maxillary palpi not sawlike, last joint hatchet-shaped, rather large ; eyes feebly sinuate ; base of thorax impressed, apex truncate; mesosternum nearly as long as intermediate coxae. Tarsi shorter than in *Xylita* and punctuation stronger.   *Zilora*, Muls.

B b. Intermediate coxae contiguous behind; third joint of antennae a little longer than fourth.

Maxillary palpi not sawlike, last joint hatchet-shaped, rather broad, furrowed on inner edge, second obconical, third triangular and a little depressed; eyes slightly sinuate; thorax truncate in front; mesosternum scarcely reaching to the half of the intermediate coxae.

Xylita, Payk.

B B. Head inclined, at least partly visible from above.

a a. Apical spines of tibiae very short, especially on anterior pair.

Maxillary palpi not sawlike, last joint hatchet-shaped, third joint transverse triangular; apex of mandibles slightly bifid; eyes sinuate in front; thorax longer than broad; scutellum very small, quadrangular; anterior coxae short, without trochantina; mesosternum as long as intermediate coxae; penultimate tarsal joint sub-bilobed; body elongate, rather flat, very finely pubescent.

Hypulus, Payk.

b b. Apical spines of tibiae long.

Last joint of maxillary palpi hatchet-shaped, elongate, furrowed on outer edge; mandibles entire at apex; eyes very narrowly and slightly emarginate; thorax transverse; scutellum rather large, elongate triangular; anterior coxae somewhat prominent, with trochantina; mesosternum much shorter than intermediate coxae; penultimate tarsal joint sub-bilobed; body elongate, broad, bare.

Melandrya, Fab.

2. Antennae ten-jointed.

Head bent downward, only slightly visible from above; last joint of maxillary palpi much longer than the rest together; apex of mandibles entire; eyes emarginate; thorax strongly transverse; scutellum rather large, elongate triangular; penultimate tarsal joint sub-bilobed; body oblong, scarcely pubescent,   *Conopalpus*, Gyll.

II. Tarsal claws toothed.

Head bent downward, only slightly visible from above; last joint of maxillary palpi hatchet-shaped, elongate, third joint triangular, nearly transverse; eyes rather large, emarginate; thorax transverse, all angles rounded; scutellum large, elongate triangular, rounded behind; elytra elongate, parallel-sided, rather flat; mesosternum reaching only to the half of the intermediate coxae; body soft, finely pubescent. Male with antennae longer than half body; posterior femora much thickened and curved, posterior tibiae ending on inner-side in a strong process and tarsal claws short, broad, trifid at apex. Female with antennae not longer than half body; posterior femora simple; posterior tibiae without apical process and tarsal claws appendiculate, their terminal division simple.   *Osphya*, Ill.

Tetratoma.

A. Elytra blackish-green or blue-black.

a. Thorax reddish-yellow.

Head and last four joints of antennae black; under-side reddish-yellow. L. 2—2¼ l. Moderately common.

T. fungorum, Fab.

b. Thorax blackish-green.

Blackish-green; antennae, palpi and legs more or less reddish. L. 2 l. Scarce.   *T. Desmaresti*, Latr.

B. Elytra reddish-yellow-brown, with black spots and bands.

Reddish-yellow-brown; apical joints of antennae, under-side and some spots on elytra black. L. 1½ l. Not uncommon.   *T. ancora*, Fab.

Orchesia.

A. Elytra unicolorous or lighter toward apex.

a. Abdomen reddish-yellow; length 2—2¼ lines.

Upper-side brown, with silky pubescence; under-side pitch-brown; abdomen and legs reddish-yellow. Base of thorax produced in middle and very slightly impressed on each side; punctuation fine and close, in wrinkles. Common.   *O. micans*, Panz.

b. Abdomen pitch-black ; length 1¼—1¾ lines.

Oblong. Under-side pitch-black ; upper-side usually rather lighter; mouth, antennae and legs reddish-pitch-brown. Base of thorax slightly bisinuate, with an impression on each side, punctuation wrinkled. Not uncommon.   *O. minor*, Walk.

B. Elytra pitch-brown, with three reddish-yellow-brown bands, the middle one strongly waved.

Head, thorax and under-side pitch-brown ; antennae and legs rust-red. Punctuation very fine. L. 2½ l. Not common.   *O. undulata*, Kr.

Hallomenus.

Brownish-yellow; thorax with two black spots ; elytra generally darker than thorax, paler at base, with very feeble striae. Punctuation very fine, wrinkled. L. 2¼—2½ l. Not uncommon.   *H. humeralis*, Panz.

Anisoxya.

Tolerably cylindrical, narrowed behind. Pitch-brown, basal and apical margins of thorax, mouth, base of antennae and legs yellow-red, pubescence gray. Punctuation fine and close. L. 1½—1¾ l. Scarce.   *A. fuscula*, Ill.

Abdera.

A. Elytra black, with light markings.

a. Middle of suture of elytra black.

I. Base of thorax reddish-yellow ; pale bands on elytra broad.

Black ; mouth, basal and apical margins of thorax reddish-yellow ; base of antennae and legs pale yellow-brown ; elytra with a light curved band before middle and a narrower one behind middle ; pubescence feeble. L. 1½—1¾ l. Scarce.   *A. quadrifasciata*, Steph.

II. Base of thorax black ; pale bands on elytra narrow.

Pitch-black, pubescent, finely punctured ; elytra glabrous, with two reddish-yellow bands, the broader one before, the narrower behind the middle ; antennae and legs dusky yellow. L. 1½—2 l. Not common. *A. bifasciata*, Marsh.

b. Middle of suture of elytra pale.

Oblong, rather shiny, convex. Black ; elytra with a not very sharply defined transverse band before middle, united along suture with a round spot before apex, pale ; base of antennae, tibiae and tarsi red-yellow. Thorax a little narrower than elytra, latter with rows of hairs ; punctuation strong and close, somewhat wrinkled. L. 1½ l. Rare.

A. triguttata, Gyll.

B. Elytra reddish-yellow-brown ; with two narrow dark bands.

Upper-side reddish-yellow-brown ; a broad straight band on thorax, scutellum and two strongly bent toothed bands on middle of elytra black ; under-side black, with prosternum and legs rust-red. Punctuation very fine and close, wrinkled ; pubescence very fine. L. 1⅔ l. Scarce.

A. flexuosa, Payk.

Serropalpus.

Elongate. Brown ; pubescence tolerably close, silky. Elytra with not very distinct striae ; punctuation very fine, wrinkled. L. 6—8 l. Doubtful if indigenous ; a specimen taken in a warehouse at Leicester in 1844.

S. striatus, Hellen.

Phloeotrya.

Pitch-brown, slightly pubescent, thick punctured ; anterior ridge of thorax and body beneath rusty-pitchy ; whole of legs and base of antennae red-yellow, apex of latter brown ; elytra sometimes dull red-yellow. L. 4—7 l. Scarce.

P. Stephensi, Duv.

Zilora.

Rust-brown ; antennae and legs lighter ; pubescence rather long, erect. Elytra parallel-sided, rounded at apex, rather strongly punctured, here and there in rows. L. 3— 3½ l. Braemar.

Z. ferruginea, Payk.

Xylita.

Pitch-black ; antennae, legs (sometimes except femora)

and often also elytra brown. Anterior coxae placed rather far from anterior margin of prosternum; punctuation close and fine, wrinkled; pubescence fine, gray. L. 2¾—4 l. Not common.   *X. laevigata,* Hellen.

Hypulus.

Pitch-brown, thorax blackish; elytra reddish-yellow-brown, with neighbourhood of scutellum, a band behind middle, a spot before middle of each and often also apex blackish; antennae and legs reddish-yellow-brown. Sides of thorax slightly rounded in front, its base with a large longitudinal impression on each side. L. 2½ l. Scarce.

H. quercinus, Payk.

Melandrya.

A. Thorax without central furrow; elytra with punctured striae throughout.

Black, upper-side often black-blue; apex of antennae and greater part of tarsi reddish-yellow. Thorax with a depression on each side at base. L. 4—6 l. Rather common.   *M. caraboides,* Lin.

B. Thorax with distinct central furrow; elytra smooth at base, but with deep striae behind.

Black, shiny; apex of antennae and tarsi reddish-yellow. Thorax deeply impressed on each side near raised lateral margin. L. 5—7 l.   Very rare. *M. canaliculata,* Fab.

Conopalpus.

A. Length 3½ lines.

Either pale reddish-yellow, with elytra paler and eyes and antennae (except base) black, or pitch-black, with base of antennae, mouth, thorax, prosternum and legs yellow. Thorax rather closely but shallowly punctured; elytra scantily punctured, basal margin reddish. Scarce.

C. testaceus, Ol.

B. Length 2¼ lines.

Similar to *C. testaceus* but with more bluish reflection on elytra, more finely and deeply punctured; thorax evenly and strongly narrowed from base forward. (? Brit.)

C. brevicollis, Kr.

Osphya.

Male. Black ; mouth, more or less of margins of thorax, base of antennae and legs (except apex of posterior femora) reddish-yellow ; disc of elytra sometimes light. Female yellow-brown ; vertex of head, two points on thorax, apex (rarely also disc) of elytra and breast black. L. 3—5 l. Scarce. *O. bipunctata*, Fab.

LAGRIIDAE.

Head with a very thick neck ; last joint of maxillary palpi strongly hatchet-shaped ; apex of mandibles bifid ; eyes strongly emarginate ; antennae gradually thickened, last joint elongate ; thorax cylindrical ; elytra dilated behind, somewhat inflated ; penultimate tarsal joint sub-bilobed. *Lagria*, Fab.

Lagria.

Black ; elytra brownish-yellow, soft : pubescence long and rough. Male with thorax as long as broad, diffusely punctured, and elytra narrow ; female with thorax broader than long, closely punctured, with central line smooth and elytra broader. L. 4—5 l. Common. *L. hirta*, Lin.

PEDILIDAE.

A. Thorax nearly semicircular, not much narrower than elytra ; pronotum separated from sides of prothorax.

Last joint of maxillary palpi hatchet-shaped ; antennae a little longer than thorax ; body strongly pubescent ; eyes emarginate. *Scraptia*, Latr.

B. Thorax nearly quadrangular, much narrower than elytra, pronotum not separated from sides of pro-thorax.

Last joint of maxillary palpi hatchet-shaped ; antennae at least as long as half body ; pubescence fine ; eyes emarginate. *Xylophilus*, Latr.

Scraptia.

A. Thorax without basal depressions.

Dusky brown ; obscure, clothed with a fine villose down ; beneath darker and more glossy; thorax short, semicircular ; elytra finely punctulated ; legs pale brown, tibiae and tarsi dull red. L. 1¾ l. Rare. *S. fusca*, Latr.

B. Thorax with basal depressions.

Pitch-black or brown ; base of antennae and legs brownish-yellow ; pubescence fine, golden-yellow. Punctuation close and fine ; thorax almost semicircular, with a small depression on each side of base. L. 1¼ l. Rare.
S. fuscula, Müll.

Xylophilus.

A. Third joint of antennae scarcely as long as second.

Reddish-yellow-brown ; with very fine, silky, whitish-gray pubescence, a spot at scutellum and a band behind middle of each elytron bare. Thorax generally with a curved impression before base ; punctuation fine. L. ¾— 1 l. Not uncommon. *X. populneus*, Fab.

B. Third joint of antennae distinctly longer than second.

a. Eyes scarcely emarginate, placed at some distance from basal margin of head.

Reddish-yellow ; elytra, meso- and metasternum and abdomen slate-coloured, former with shoulders and hinder portion reddish-yellow. Thorax scarcely as long as broad, with a transverse impression before base (often reduced to two depressions) ; punctuation fine. L. ¾—1 l. Rare.
X. neglectus, Duv.

b. Eyes distinctly emarginate, reaching to basal margin of head.

Head and thorax black, latter sometimes paler ; elytra, antennae and legs reddish-yellow ; under-side black or brown, more or less yellowish in parts. Thorax broader than long, with a transverse furrow (sometimes indistinct) before base ; punctuation distinct. L. ¾—1 l. Not uncommon. *X. pygmaeus*, De G.

ANTHICIDAE.

A. Thorax with a horn in front.

Head oblong ; last joint of maxillary palpi hatchet-shaped; mandibles broad, straight, apex abruptly curved and two-toothed. *Notoxus*, Geoffr.

B. Thorax without horn.

Head rounded or quadrangular; last joint of maxillary palpi hatchet-shaped; mandibles broad, curved, bifid at apex. *Anthicus*, Payk.

Notoxus.

Reddish-yellow-brown; head dark; elytra with a small spot at scutellum, another (sometimes absent) at side before middle, and a large crescent-shaped one behind middle (often united with scutellary or lateral spot) blackish. Horn of thorax with four or five rounded teeth at sides; elytra of male somewhat truncate, with a slight elevation. L. 1½— 2 l. Common. *N. monoceros*, Lin.

Anthicus.

A. Head rounded at base.

a. Front of thorax narrower than head.

Pitch-black; head black; elytra often lighter at shoulder and also before apex, sometimes entirely light brown; base of antennae and legs (with or without femora) red-brown; pubescence rather close, gray. Head completely rounded behind; thorax oblong heart-shaped, strongly dilated in front, constricted before base. L. 1—1¼ l. Rather common. *A. humilis*, Germ.

b. Front of thorax as broad as head.

Black, with gray pubescence; antennae short, black-brown; femora pitchy, tibiae and tarsi reddish. Head diffusely and finely punctured; thorax as broad in front as head, strongly narrowed behind, constricted before base, very closely and finely punctured, dull; elytra shiny, diffusely and not very strongly punctured, somewhat wrinkled transversely. Punctuation finer and closer than in *A. humilis*. L. 1—1⅛ l. Not common.
A. salinus, Crotch.

B. Head truncate, or nearly so, at base.

a. Apex of elytra pale.

Pale brown-yellow; abdomen and a spot on disc of each

elytron, somewhat behind middle, black; pubescence fine, gray; punctuation extremely close. Black spot on elytra variable, sometimes absent, at others enlarged into a streak and joined at suture to that on other elytron. L. 2—2¼ l. Rare. *A. bimaculatus*, Ill.

b. Apex of elytra dark.

I. Elytra yellow-red or red-brown at base, becoming gradually darker in or before middle.

1. Elytra flat, impressed near base, shoulders prominent.

Black-brown, shiny; thorax, front part of elytra, antennae and legs rust-red; pubescence feeble. Base of head with a longitudinal furrow; thorax with two small elevations in middle of front. L. 1¼—1½ l. Common. *A. floralis*, Lin.

The variety *quisquiliarius*, Th. is less shiny, thorax without tubercles in front and punctuation much closer.

2. Elytra rather convex, not impressed near base, shoulders not prominent.

Dark red-brown; head pitch-brown; elytra pitch-brown from middle backward; antennae, tibiae and tarsi red-yellow. Head broader than long, base without furrow, eyes very prominent, hemispherical; thorax broader in front than head (without eyes), much narrowed behind; elytra oval, broadest in middle; posterior tibiae of male much dilated at apex; pubescence distinct, gray; punctuation strong and rather close throughout. L. 1½ l. Rather common. *A. instabilis*, Schmidt.

II. Elytra unicolorous black or brown.

1. Thorax short; femora red-brown.

A A. Body dark brown; shoulders of elytra not prominent.

Elongate, narrow. Dark brown, with ochre-yellow pubescence; base of head and thorax red, disc of latter dark; antennae and legs pale red-brown. Thorax ovate; punctuation close and rather coarse. L. 1 l. Not uncommon. *A. angustatus*, Curt.

B B. Body leaden-black; shoulders of elytra prominent.

Leaden-black, dull, closely covered with shiny gray

pubescence; antennae and legs lurid red-yellow or almost entirely pitch-brown. Head broad, truncate behind, strongly and closely punctured, almost in wrinkles, central line impunctate; thorax short, closely and not very strongly punctured; elytra much broader than thorax, truncate at base, convex, short ovate at sides on apical third-part, punctuation tolerably strong, almost confluent. Broader than *A. angustatus*, with antennae shorter and stouter, thorax broader, elytra much wider and shoulders more prominent. L. 1¼ l. Scarce. *A. scoticus*, Rye.

2. Thorax long; femora black.

Black; antennae and tibiae brown; tarsi reddish; pubescence fine. Head as long as broad, eyes small; thorax longer than broad, sides nearly straight from before middle to base, nearly as broad as head in front, much narrowed behind; elytra oblong-ovate; punctuation rather close, that of head and thorax fine, that of elytra stronger. L. 1¼ l. Rather common. *A. Schaumi*, Woll.

III. Elytra black, with well-defined yellowish-red markings.

Black; a large spot on elytra near base and an oblique band (usually dilated at suture before and behind) behind middle yellowish-red; antennae black; tibiae sometimes brown; tarsi yellow-brown; pubescence fine and rather close, gray. L. 1⅓—1½ l. Common.

A. antherinus, Lin.

PYROCHROIDAE.

Last joint of maxillary palpi hatchet-shaped, narrow and acute at apex; eyes widely emarginate, placed much apart above. *Pyrochroa*, Geoffr.

Pyrochroa.

A. Head black.

Black; thorax and elytra scarlet. Forehead with a quadrangular impression, rounded behind and in male sharply defined. L. 7—8 l. Moderately common.

P. coccinea, Lin.

B. Head red.

a. Thorax blood-red; length 4½—5½ lines.

Elytra red : breast and abdomen black. Forehead with a deep crescent-shaped impression ; thorax with a fine central furrow ; antennae of male sawlike. Common.

P. serraticornis, Scop.

b. Thorax yellow-red, with dark disc ; length 3½ lines.

Elytra yellow-red ; breast and abdomen black. Forehead uneven ; each elytron with two slightly raised lines ; antennae of male comblike. Rare. *P. pectinicornis,* Lin.

MORDELLIDAE.

A. Pygidium prolonged into a conical process.

Posterior coxae as long as metasternum ; tarsal claws more or less toothed ; eyes entire.

a. Scutellum large, transversely rectangular.

Similar to *Mordella,* but with antennae sawlike from fifth joint, gradually decreasing ; intermediate tibiae shorter than the first four joints of their tarsi.

Tomoxia, Cost.

b. Scutellum not large, transversely triangular or rounded.

I. Antennae obtusely sawlike toward apex ; episterna of metathorax elongate triangular.

Last joint of maxillary palpi hatchet-shaped ; antennae simple or sawlike from fourth or fifth joint ; thorax transverse ; penultimate joint of front pairs of tarsi emarginate or hollowed out above.   *Mordella,* Lin.

II. Antennae almost threadlike ; episterna of metathorax elongate linear.

Similar to *Mordella,* but with thorax at least as long as broad.   *Mordellistena,* Cost.

B. Pygidium curvilinear triangular, not prolonged.

Last joint of maxillary palpi hatchet-shaped, more or less elongate ; eyes generally emarginate ; antennae nearly threadlike or slightly thickened at apex ; thorax strongly transverse ; fourth joint of anterior tarsi very small, received in third, which is bilobed.   *Anaspis,* Geoffr.

Tomoxia.

Black, with close, silky, grayish-yellow pubescence, placed more closely on base of elytra and on a spot behind middle of each; base of antennae reddish-yellow. Elytra finely and scantily punctured; anal process sharply pointed. L. 3¼—4 l. Not common. *T. biguttata*, Gyll.

Mordella.

A. Pubescence of upper-side forming spots or bands in places.

Black, with rather close, silky, gray pubescence, placed more closely on an oblique spot at shoulder and a straight band (interrupted at sides and suture) behind middle of each elytron; antennae sawlike, with base yellow-brown. Anal segment produced into a long, sharp point. L. 2½ l. Not common. *M. fasciata*, Fab.

B. Pubescence of upper-side evenly distributed.

Black, with rather thick, silky, closely-lying pubescence. Thorax much broader than long, its process over scutellum truncate or slightly emarginate; elytra rather more than double as long as broad at base, much narrowed behind; anal spine as long as abdomen; antennae sawlike. L. 2 —2¼ l. Very rare. *M. aculeata*, Lin.

Mordellistena.

A. Thorax wholly or partly yellow.

a. Abdomen reddish-yellow.

Black, with fine, silky, gray pubescence; mouth, base of antennae, anterior legs and abdomen reddish-yellow, hinder legs dark, but their tarsi partly reddish; thorax of male black, with base brown-yellow, that of female red. L. 2— 2½ l. Not common. *M. abdominalis*, Fab.

b. Abdomen black.

I. Antennae darker at apex than at base.

1. Head black.

Black, with fine, silky, gray pubescence⁻; mouth, base of antennae, sides of thorax, a spot on shoulders of elytra and the anterior legs reddish-yellow; hinder femora narrowly

at base and apex, hinder tibiae broadly at base yellow-
brown. L. 1½—2 l. Not common. *M. humeralis*, Lin.

2. Head reddish-yellow-brown.

Reddish-yellow-brown; eyes black; apex of elytra and
of abdomen often brown; apex of antennae brown. Thorax
broader than long, narrowed in front, its process over scutellum truncate; elytra almost three times as long as
together broad at base, not much narrowed behind; antennae threadlike. L. 1¾ l. Not common.

M. brunnea, Fab.

II. Antennae entirely reddish-yellow.

Black; mouth, antennae, a more or less distinct spot at
anterior angles of thorax (or almost the whole of its side
margin), a humeral spot on elytra, gradually narrowed and
extending beyond middle and the legs reddish-yellow; posterior femora sometimes dark in middle; pubescence fine,
gray. Antennae very feebly sawlike; anal process long,
pointed. L. 1⅓—1¾ l. Not common.

M. lateralis, Ol.

B. Thorax entirely black.

a. Process of thorax over scutellum slightly emarginate.

I. Anal spine long.

Black, with moderately close, silky pubescence. Thorax
as long as broad, not narrowed in front, base strongly bisinuate, posterior angles acute; elytra three times as long
as together broad at base, parallel-sided nearly to apex;
antennae scarcely sawlike. L. 1½ l. Rather common.

M. pumila, Gyll.

II. Anal spine short.

Black, not very shiny, with silky pubescence. Thorax
with posterior angles right angles, a little shorter than in
M. pumila, base less strongly sinuate. L. 1½ l. Scarce.

M. brevicauda, Boh.

b. Process of thorax over scutellum broadly rounded.

Black, with silky pubescence; mouth and base of antennae pitch-brown. Thorax about as broad as long, base
slightly bisinuate, posterior angles almost obtuse; anal

spine short. L. 1¼ l. Not uncommon.

M. inaequalis, Muls.

Anaspis.

A. Thorax black.

a. Elytra entirely black.

I. Appendices of third abdominal segment of male curved at apex, fifth segment foveolate.

Black, with very fine brown pubescence ; base of antennae, mouth and anterior legs rust-red. Posterior angles of thorax right angles; elytra with very fine transverse scratches. L. 1¼ l. Common. *A. frontalis*, Lin.

II. Appendices of third abdominal segment of male linear, reaching fifth segment, which is straight.

Black, shiny ; base of antennae red-yellow ; mouth and legs brown, anterior pair dark rust-red. Posterior angles of thorax nearly acute. More distinctly strigose than *A. frontalis*, antennae with second joint longer, fourth narrower and a little shorter than fifth, last joint of palpi less broadly hatchet-shaped. L. 1¼ l. Not common.

A. rufilabris, Gyll.

III. Appendices of third abdominal segment of male reaching beyond fifth segment, which is bilobed.

Black, pubescent ; base of antennae, mouth and front pairs of legs yellow ; posterior legs generally black or brown. Antennae thickened from seventh joint, second joint at least half as large as third, latter and fourth longest, nearly equal, sixth to tenth longer than broad. L. 1¼—1⅔ l. Common. *A. forcipata*, Muls.

b. Elytra black, with a light spot at shoulder.

Black ; base of antennae, a spot at shoulder of each elytron and usually also anterior tibiae red-yellow ; pubescence fine, gray. L. 1—1¼ l. Common. *A. fasciata*, Forst.

B. Thorax red or yellow-brown.

a. Elytra black.

I. Hinder half of head black.

Similar to *A. thoracica* but with hinder half of head

black, legs yellow, posterior femora rather dark and with pubescence somewhat closer. L. 1 l. Common.

A. ruficollis, Fab.

II. Whole of head reddish-yellow.

Oblong. Black or dark brown ; base of antennae, whole of head, thorax and front pairs of legs reddish-yellow, posterior legs brown, with femora darker ; pubescence fine, gray. L. 1⅓ l. Moderately common.

A. thoracica, Lin

b. Elytra entirely yellow.

Red-yellow, slightly pubescent, shiny ; eyes and apex of antennae black. L. 1½ l. Not uncommon.

A. subtestacea, Steph.

c. Elytra yellow-brown, with dark spots.

Elongate. Yellow-brown ; apex of antennae, eyes, breast, and abdomen black ; elytra with two dark spots on each, and a third common one behind middle ; pubescence rather close, gray. L. 1¼ l. Common.

A. melanopa, Forst.

RHIPIPHORIDAE.

Mentum slender, confused with ligula ; labial palpi apparently consisting of a single elongate, spindle-shaped joint ; last joint of maxillary palpi very long, curved, truncate at apex ; maxillary lobes rudimentary; mandibles not toothed on inner-side ; vertex of head depressed, scarcely more prominent than anterior margin of thorax, eyes oval ; antennae of male with two, of female with one process to each joint from third onward ; thorax longer than broad ; elytra imperfectly covering abdomen, gaping ; intermediate coxae nearly contiguous. *Metoecus*, Gerst.

Metoecus.

Black ; sides of thorax and abdomen yellow-red ; elytra of male wholly or partly yellow. Sides of thorax straight and convergent forward, disc with a deep central furrow, abbreviated in front, base produced in middle, posterior angles prominent. L. 3¼—4½ l. Not common.

M. paradoxus, Lin.

MELOIDAE.

A. Metasternum very short; intermediate coxae partly covering posterior.

Last joint of maxillary palpi cylindrical, depressed and obtuse at apex; apex of mandibles truncate or slightly emarginate; eyes slightly emarginate; second joint of antennae very short, last joint generally elongate, cylindrical and pointed at apex, scutellum absent; elytra overlapping at suture, divergent, more or less abbreviated; body apterous. *Meloë*, Lin.

B. Metasternum elongate; intermediate coxae distant from posterior.

a. Elytra entirely covering abdomen, not divergent.

Antennae not very short, not clubbed; eyes transverse, very slightly emarginate; penultimate tarsal joint more or less cylindrical, not bilobed; claws not toothed; body winged. *Cantharis*, Geoffr.

b. Elytra abbreviated, narrowed behind, divergent.

Antennae threadlike; upper division of tarsal claws generally simple, sometimes (in same species) slightly toothed; body winged. *Sitaris*, Latr.

Meloë.

A. Thorax as long as, or longer than broad.

a. Head and thorax closely and coarsely punctured.

Black, with blue or violet reflection; antennae and legs black-blue. Thorax without impressions and with almost straight posterior margin; elytra coarsely wrinkled, covering in male nearly whole of abdomen. L. 6—14 l. Common. *M. proscarabaeus*, Lin.

b. Head and thorax with only scattered punctures.

Dark blue, shiny. Thorax with a transverse impression before the rather deeply emarginate base; elytra shagreened. L. 6—10 l. Rather common. *M. violaceus*, Marsh.

B. Thorax broader than long.

a. Thorax with rather fine, scattered punctures.

Upper-side black, with blue reflection. Thorax not much broader than long, slightly dilated before middle, base emarginate, disc with central furrow; elytra with scattered punctures, rather larger and more shallow than those of thorax. L. 8 l. Rare.   *M. autumnalis*, Ol.

b. Thorax strongly punctured.

I. Head not, or not much broader and larger than thorax.

1. Body black.

Black, shiny; elytra with blue reflection; thorax flat, with a short, fine central furrow and outward-pointed anterior angles. L. 14—18 l. Not common.

M. cicatricosus, Leach.

2. Body chiefly greenish.

Upper-side dirty metallic-green; head and thorax with purple margins; each abdominal segment with a large coppery spot above; under-side green; base of each ventral segment coppery. L. 9—15 l. Rare.

M. variegatus, Don.

II. Head considerably broader and larger than thorax.

1. Thorax closely punctured, with a deep curved impression before the emarginate base.

Black, dull. Head and thorax closely and very coarsely punctured, with central furrow, and latter broadest rather before middle. L. 5—7 l. Rare.   *M. rugosus*, Marsh.

2. Thorax scantily and simply punctured; posterior margin straight.

Blue-black. Antennae beadlike; thorax rounded at sides. L. 5—10 l. Rare.   *M. brevicollis*, Panz.

Cantharis.

Bright green; antennae and legs dark. Head and thorax with fine scattered punctures; elytra finely and closely punctured in wrinkles, with some slightly raised lines; head with strong central furrow. L. 6—9 l. Not common.

C. vesicatoria, Lin.

Sitaris.

Black; elytra bluish-black, with base yellowish. Thorax

coarsely punctured ; elytra narrowed from shoulders, separately pointed at apex, finely punctured. L. 5—6 l. Scarce.

S. muralis, Forst.

OEDEMERIDAE.

A. Only penultimate joint on all tarsi with thickly matted
hairs beneath.

a. Head produced in front.

I. Eyes round.

Antennae eleven-jointed ; last joint of maxillary palpi
oblong conical ; posterior femora of male nearly always
much thickened : all tibiae with two apical spines.

Oedemera, Ol.

II. Eyes kidney-shaped.

Antennae eleven-jointed ; last joint of maxillary palpi
long, somewhat dilated toward the obliquely truncate apex ;
eyes strongly granulate ; posterior femora of male much
thickened ; all tibiae with two apical spines.

Oncomera, Steph.

b. Head not produced in front.

Antennae eleven-jointed ; eyes slightly emarginate,
finely granulate ; last joint of maxillary palpi knife-shaped
in male, hatchet-shaped in female ; legs simple, all tibiae
with two apical spines. *Asclera*, Schmidt.

B. First four joints of anterior tarsi, middle joints of intermediate pair and penultimate joint of posterior pair with
thickly matted hairs beneath.

Antennae twelve-jointed in male, eleven-jointed in
female ; eyes kidney-shaped ; legs simple, anterior tibiae
with one, hinder pairs with two apical spines.

Nacerdes, Schmidt.

Oedemera.

A. Elytra strongly pointed at apex.

Green or blue, shiny ; under-side of base of antennae
and the base of anterior tibiae yellow. Thorax longer than
broad ; elytra with three slightly raised longitudinal lines ;
posterior femora of male very strongly thickened. L. 4½
—5 l. Common. *O. nobilis*, Scop.

B. Elytra only slightly narrowed toward apex.

Dull green. Thorax scarcely longer than broad; elytra with three sharply raised longitudinal lines; posterior femora of male scarcely thickened. L. 3 l. Common.

O. lurida, Marsh.

Oncomera.

Pale yellow-brown; forehead, sides of thorax, base of abdomen and a ring before apex of femora brown. Elytra gaping at suture, separately rounded at apex, with three or four more or less distinct raised lines. L. 6—7 l. Not uncommon.   *O. femorata,* Fab.

Asclera.

A. Thorax blue or blue-green.

Blue or blue-green; antennae black. Thorax with feeble impressions; disc of each elytron with three raised longitudinal lines. L. 3½—4 l. Rather common.

A. coerulea, Lin.

B. Thorax reddish-yellow.

Dark green; thorax reddish-yellow; under-side of base of antennae yellow. Thorax with three depressions on disc: elytra very finely punctured, each with three fine raised lines. L. 4—5 l. Scarce.

A. sanguinicollis, Fab.

Nacerdes.

Head and thorax reddish-yellow; elytra yellow, with apex black; under-side pitch-black, legs brown-yellow. In male the thorax heart-shaped, with a black spot at sides, and the apex of last abdominal segment deeply triangularly excised, in female the thorax almost quadrangular and the last abdominal segment yellow, rounded at apex. L. 5—7 l. Not uncommon.   *N. melanura,* Lin.

ABERRANT COLEOPTERA.

Male. Mouth parts in a state of atrophy, except mandibles and two palpi; head prominent, vertical, transverse,
prolonged on sides; eyes prominent, very strongly granulate, carried by lateral projections of head; antennae
inserted on inner-side of base of latter, with from four to
seven joints, forked; prothorax and mesothorax very
short, soldered together, metathorax extremely large;
elytra membranous or leathery, very small and narrow;
wings very large, fan-shaped; legs moderately long, feeble;
coxae short, nearly globular, the front pairs placed a little
apart at base, the posterior pair almost contiguous; trochantina very elongate and prominent; tibiae without
apical spines; tarsi with from two to four joints, with
membranous prominences beneath, without claws; abdomen with from seven to nine segments.

Female. Apterous and larva-like.

Parasitic on certain *Hymenoptera*; female imago partly
enclosed between abdominal segments of its host throughout its life; male imago living at longest a single day.

Stylopidae.

STYLOPIDAE.

A. Tarsi with four joints.

Antennae with six joints, first elongate, second and third
very short, latter with a long process, fourth much larger

than previous two together, last two gradually smaller; eyes with numerous facets; elytra elongate oval, narrowed at base; wings nearly as broad as long; post-scutellum elongate, rounded at apex; posterior trochanters elongate.   *Stylops*, Kirby.

B. Tarsi with two joints.

Antennae slender, with five joints, first two short, third transverse, with a long process, fifth half as long again as fourth; eyes with only about twenty facets; elytra narrow, attenuate at base; wings longer than broad; post-scutellum oval; posterior trochanters short.   *Elenchus*, Curt.

C. Tarsi with three joints.

Antennae with seven joints, first two larger than rest, latter with appendages on outer-side, these appendages somewhat oval, decreasing in length until last joint, which is inserted at base of penultimate; eyes strongly granulate; elytra very slender at base and ending in an oval dilatation; wings longer than broad; post-scutellum very elongate, with a long deep furrow at base.
                                            Halictophagus, Curt.

Stylops.

Deep black; legs brown; wings milky-white, nervures blackish. L. 1½ l. Rare.   *S. melittae*, Kirby.

Elenchus.

Deep black; eyes nearly sessile; antennae slender, pitchy, their processes linear; wings blackish. L. ½ l. Rare.   *E. tenuicornis*, Kirby.

Halictophagus.

Black and slightly glossy, clothed with a brown-velvety pubescence; antennae and legs dull brownish-ochre; wings slightly tinged with brownish-ochre and obscurely iridescent; nervures brown; apex of tarsal joints and of abdomen ochre-yellow. L. ½ l. Rare.   *H. Curtisi*, Curt.

INDEX.

VOLUME II.

B B